Nelson Advanced Science

Make the Grade

AS
Biology

with Human Biology

**John Adds, Alan Clamp, Martin Furness-Smith,
Erica Larkcom, Ruth Miller**

Nelson Thornes
Delta Place
27 Bath Road
Cheltenham
GL53 7TH
United Kingdom

First published in 2001

ISBN 0-17-448279-5

01 02 03 04 / 10 9 8 7 6 5 4 3 2

Typesetting and illustrations by Hardlines, Charlbury, Oxford
Printed in China by L.Rex Printing Co., Ltd

Acknowledgements

The authors wish to make the following acknowledgements:
Thank you to Helen: For explaining predation by bats, athletic constraints in elephants and the classification of mammals. [AC.]

The practice questions and mark schemes are based upon existing Edexcel Foundation questions. The questions should be seen as indicative of the knowledge required rather than of the layout of the test.

In past examination questions, where magnification has been given beside a diagram or photograph in the original question, this magnification has been adjusted for the size of the diagram or photograph as it appears in this book.

Every effort has been made to trace all the copyright holders, but where this has not been possible the publisher will be pleased to make any necessary arrangements at the first opportunity.

Contents

Introduction		iv
How to make the grade		vi
Unit 1	**Molecules and cells**	1
Topic 1	Biological molecules	2
Topic 2	Enzymes	7
Topic 3	Cellular organisation	10
Topic 4	The cell cycle	14
Assessment questions		17
Unit 2B	**Exchange, transport and reproduction**	20
Topic 1B	Exchanges with the environment	20
Topic 2B	Transport systems	24
	Transport in flowering plants	24
	Transport in mammals	28
Topic 3B	Adaptations to the environment	31
Topic 4B	Sexual reproduction	34
Assessment questions		39
Unit 2H	**Exchange, transport and reproduction in humans**	44
Topic 1H	Exchanges with the environment	44
Topic 2H	Transport of materials	47
Topic 3H	Human ecology	50
	Extremes of temperature	50
	High altitude	53
Topic 4H	Human reproduction and development	57
Assessment questions		60
Unit 3	**Energy and the environment**	61
Part a	Topics	62
Topic 1	Modes of nutrition	62
Topic 2, 3	Ecosystems and energy flow	64
Topic 4	Recycling of nutrients	67
Topic 5	Energy resources	70
Topic 6	Human influences on the environment	73
	Deforestation and desertification	74
	Atmospheric and water pollution	75
Assessment questions		81
Part b	The individual investigation	84
Answers		86
Answers to assessment questions		104
Index		107

Introduction

How to use this book

This book has been written to be used alongside the two Nelson Advanced Science (NAS) Biology student books for Advanced Subsidiary (AS) and the course guide *Tools, Techniques and Assessment in Biology* (see back cover for details). It aims to help you to develop your study skills, make your learning more effective and to give you help with your revision. Now that courses are covered unit by unit, you could have unit tests at different stages of your course, so you need to be prepared right from the beginning. The early development of good study skills and sensible organisation of your notes (and your time) could greatly improve your chances of success. 'How to make the grade', which follows on page vi, contains useful advice to help you.

The contents of the book are arranged in the order of the units of the Edexcel specification (syllabus), so you should be able to find your way around fairly easily. To help you, the chapter numbers correspond to the AS units. You may notice that there appear to be two versions of Unit 2: Unit 2B is for students studying AS Biology and Unit 2H is for those studying AS Human Biology. Make sure you read the correct section!

Study these introductory chapters carefully before you tackle the rest of the book. They are full of useful advice.

Throughout the book you will find comments in the margins, like the one next to this paragraph. These comments fall into two categories:

- *Helpful hints* provide advice, guidance and warnings related to studying Biology;

- *Examiner's comments* are found in the answer sections for each unit and will explain the answers given, or point out common mistakes.

The structure of this book

Each unit is introduced and is then broken down into major topics, which are identical to the topics listed in the Edexcel specification. Within each topic you will find:

- an **introduction**
- a **checklist of things to know and understand**
- questions and activities to **test your knowledge and understanding** (answers are provided at the end of the book)
- **practice questions** similar to those you may get in unit tests (answers are provided at the end of the book).

Those practical activities contained within a unit are shown in italics (similar to the specification) in the 'Checklist of things to know and understand'. These are usually accompanied by a *Practical work – Helpful hints* box on how to deal with questions on the practicals within the unit tests. In addition, the chapter that covers Unit 3 contains information on the unit tests and also the practical (coursework) assessment with advice on how this should be carried out and presented.

At the end of each unit there are some **assessment questions** to try as you prepare for your unit test. A wide range of the different types of test questions is included in this book, together with answers and comments from examiners, explaining what a good answer should contain.

You will find a 'Physical Science Background' appendix at the end of Nelson Advanced Science Molecules and Cells. This describes the physical science, particularly physical chemistry, that will be useful for you to be familiar with while you study Biology or Human Biology. It can also be found in the Science area of the Nelson Thornes web site.

Note: This book covers the three units of the Advanced Subsidiary course. If you go on to study for the full Advanced GCE qualification, *Make the Grade in A2 Biology and Human Biology* carries on from this book. It covers Units 4, 5 and 6 of the Edexcel Advanced GCE specification and will be valuable during your revision. (See back cover for details.)

How to make the grade

What is meant by Advanced Subsidiary (AS)?

Advanced Subsidiary (AS) is a new one-year advanced course which replaces the Advanced Supplementary examination from September 2000. It also represents the first half of an Advanced GCE qualification, although you do not automatically have to carry on to the second half (A2) at the end of the course, as AS is a qualification in its own right. The standard of the Advanced Subsidiary (AS) GCE falls halfway between GCSE and Advanced (A) GCE.

The structure of the whole Advanced GCE course is shown below.

ADVANCED GCE (Six units over two years: AS year + A2 year)

START → Unit 1 + Unit 2 + Unit 3 →   Unit 4 + Unit 5 + Unit 6 → FINISH
        AS year (50 % of marks)   A2 year (50 % of marks)

Table 1 *Summary of the Edexcel specification content for AS Biology and Human Biology*

Unit 1 Molecules and cells This unit includes: cells and organelles; molecules; enzymes; chromosomes and the genetic code; protein synthesis.	
Unit 2B Exchange, transport and reproduction This unit includes: exchanges with the environment in a number of organisms; transport in flowering plants and mammals; adaptations to the environment in a range of species; sexual reproduction in plants and humans.	**Unit 2H Exchange, transport and reproduction in humans** This unit includes: exchanges with the environment in humans; transport in humans; adaptations to the environment in humans; human reproduction and development.
Unit 3 Energy and the environment *Part a* This half unit includes: modes of nutrition; energy flow through ecosystems; recycling of nutrients; energy resources; human influences on the environment. *Part b* This half unit represents the individual investigation **T1**, which is assessed by teachers and moderated by Edexcel.	
Note: 2B refers to Biology and 2H to Human Biology.	

Starting out in AS Biology and Human Biology

Achieving good grades in any subject involves organisation, learning and the ability to apply your knowledge and understanding. When you decide to study AS Biology or Human Biology, it is often assumed that you have already been successful at GCSE (Key Stage 4), or an equivalent course, and that you have some background knowledge of Mathematics, Physics and Chemistry. However, some students start AS Biology courses without this background and can still be successful. The important thing is to make a good start to your course.

How will the course be different from GCSE?

When studying AS courses, you will be expected to know and understand much more factual information than was needed for GCSE. Even though you might have studied a topic at GCSE, more detail is added during the AS course. There is more emphasis on the application of your knowledge to new situations; you should be able to analyse biological data and suggest explanations based on your knowledge. You also need to plan and carry out your own investigations, assessing how good your methods were and whether you could have improved the reliability of your results. It is very important to be able to communicate your knowledge clearly and effectively, using the appropriate scientific terms. These skills are built upon if you decide to study the A2 part of the course (leading to an Advanced GCE qualification).

The specification

You can download a copy of the specification from Edexcel's website. Visit http://www.edexcel.org.uk

It is a good idea to have a close look at the specification (syllabus). Your teacher should have one, or you can obtain your own copy from Edexcel. In addition to describing the content of the units, the specification provides information about written tests and coursework assessment.

It is important for you to understand the key terms used in the specification, as defined below.

Recall	Identify and revise biological knowledge gained from previous studies of Biology, including other units within the specification, e.g. *recall* the structure of DNA (Unit 1).
Know	Be able to state facts, or describe structures and processes, from material within the unit, e.g. *know* the internal structure of a dicotyledonous leaf (Unit 2).
Understand	Know the underlying principles and be able to apply this knowledge to new situations.
Appreciate	Be aware of the importance of biological information, without having a detailed knowledge of the underlying principles.
Discuss	Give a balanced, reasoned and objective review of a particular topic.
Describe	Provide an accurate account of the main points (an explanation is not necessary).
Explain	Give reasons, with reference to biological facts and theories.

The Advanced Subsidiary (AS) unit tests

It is useful for you to read the examiners' reports and published mark schemes from previous unit tests (these are available from Edexcel). These documents will show you the depth of knowledge that examiners are looking for in answers, as well as pointing out common mistakes and providing advice on how to achieve good grades in the tests.

The assessment scheme

AS is made up of the first three units of the Advanced GCE course. In the Edexcel specification, Units 1 and 2 are assessed by written papers, each 1 hour and 20 minutes long. Unit 3 has a shorter written paper of 1 hour, and a coursework element (T1), which is worth 15 per cent of the final AS mark.

A summary of the unit test papers is given in Table 2 on page viii.

Table 2 *Summary of the unit test papers for AS Biology and AS Human Biology*

Unit test	Mark allocation	Time allowed for the written test	Types of questions
1	70	1 hour 20 minutes	about nine structured questions, including at least one free prose question, worth 4 to 12 marks each
2B or 2H	70	1 hour 20 minutes	about nine structured questions, including at least one free prose question, worth 4 to 12 marks each
3 *Part a*	38	1 hour	about three structured questions, one short and two long with stimulus material, worth 6 to 20 marks each
3 *Part b*	32	not applicable	practical coursework assessment

Notes: B or H in the unit test column refers to Biology or Human Biology.
All of the questions in the written test for each unit are compulsory.

Unit 1

In the Unit 1 test, the specification content is the same for Biology and Human Biology, so this is a common paper. The shorter questions test mainly your knowledge and understanding of the specification content, but the longer questions may also require you to show that you can interpret data presented to you in the form of tables or graphs. One of the questions will be a 'free prose' question and advice on how to tackle these is given in this introduction (see page xiv).

Unit 2

The questions in the Unit 2 test are similar in format to those in the Unit 1 test. However, the specification is slightly different for Biology (2B) and Human Biology (2H), and this difference is reflected in the layout of the test. The first section contains questions which test the common parts of the specification, such as the structure and roles of arteries and veins, but there are then separate sections for the Biology candidates and the Human Biology candidates. It is important to make sure that you answer the correct section of the paper. You may think that you can answer some of the questions in the Biology section if you are a Human Biology candidate (or *vice versa*), but it is unlikely that your answers would be sufficiently detailed to enable you to gain a high mark.

Unit 3

Unit 3 has two parts, a **written test** and an **individual investigation**. The written test is quite short, with only three questions. However, two of these questions present you with material to read and data to interpret. You need to read through this stimulus material very carefully and then make sure that you produce a detailed and relevant answer. It should be quite easy to judge the level of detail required by looking at the mark allocations for each part of the question (as for Unit 1 and Unit 2 tests). The main difference in Unit 3 is that you may not have come across the data before and you will need to apply your biological knowledge to unfamiliar situations. Often there may not be a single right answer, but you may be asked to suggest reasons and explanations based on your knowledge of the specification. (See Unit 3, Part b for advice on the practical investigation, page 84.)

Study skills

Students who are interested in Biology and who have qualifications in the subject already will still need to develop good study skills if they are to be successful. This section provides advice and guidance on how to study Biology, as well as some tips on effective revision.

Organising your notes

Biology students usually accumulate a large quantity of notes and it is useful to keep this information in an organised manner. It is advisable to file your notes in specification (syllabus) order, using a consistent series of informative headings, as illustrated below.

> ### UNIT 1
> ### Molecules and Cells
> **Enzymes**
> Enzymes are protein molecules which ...

After the lessons, it is a good idea to check your notes using your textbook(s) and fill in any gaps in the information. Make sure you go back and ask the teacher if you are unsure about anything, especially if you find conflicting information in your class notes and textbook.

When making notes from a textbook, it is useful to identify the main idea you are focusing on, such as 'enzymes'. Notes made on the topic should always relate to this key idea. You should try to get into the habit of recording the main points; details can be added later, if required. Good notes should always be clear and concise. You may find it useful to use standard abbreviations of biological terms, such as *DNA* and *ATP*, and also to use 'shorthand' terms, such as *temp* (temperature), *phs* (photosynthesis) and *conc* (concentration). However, it is not a good idea to use 'home-made' abbreviations in the unit tests – the examiner may not understand them!

The presentation of notes is important. You could try organising your notes under headings and sub-headings, with key points highlighted using capitals, italics or colour. Numbered lists are useful, as are tables and diagrams. This book contains a number of techniques used for presenting information in an interesting and an accessible manner, such as concept maps (see page xi). Try these methods to develop the habit of effective note-making.

Organising your time

It is vital to know when you will be sitting the unit tests. This information will enable you to plan your revision. Try to make a revision timetable. This should allow enough time to cover all the material, but also be realistic. It is useful to leave some time at the end of the timetable, just before the unit tests, to catch up on time lost, for example through illness. You may not be able to work for very long at a single session – probably no more than one hour – without a short break of 10–15 minutes. It is also useful to use spare moments, such as when waiting for a bus or train, to do short snippets of revision (perhaps using revision cards, see page xii). These 'odd minutes' can add up to many hours.

Make studying a habit, not a chore

Good study skills can become a habit. Try to use the 'Daily tasks', 'Weekly tasks' and 'End of topic tasks' outlined below.

Daily tasks

After each lesson, check that your notes are complete by spending between 10 and 15 minutes looking through them. If there is something you do not understand:

- Read the relevant part in this book or your textbook and, if necessary, add to your notes so that they will be clear when you read them again.
- Discuss the problem with another student.
- If you still have difficulty, ask your teacher as soon as you can.

Weekly tasks

- Go through your notes and highlight important points.
- Read through the relevant parts of this book and make notes and / or highlight important points.
- Check that you have completed all your homework assignments.

End of topic tasks

To revise a topic:

- Work carefully through your notes, with a copy of the specification (syllabus) to make sure you have not missed anything out.
- Summarise your notes to the bare essentials, using the tips given on pages xi to xii.
- Work through the topic in this book, discussing any difficulties with your teachers or other students.

Improving your memory

There are several things you can do to improve the effectiveness of your memory for biological information. Organising the material will help, especially if you use topic headings, numbered lists and diagrams. Repeatedly reviewing your notes will also be useful, as will discussing topics with teachers and other students. Finally, using mnemonics (memory aids) can make a big difference, such as **A**rteries carry blood **A**way from the heart.

Effective revision

This book provides plenty of advice and guidance on effective revision, and specific advice on techniques to use in the unit tests are given below. The following points are useful when revising.

- Try revising in a quiet room, sitting at a desk or table, with no distractions.
- Take regular breaks.
- Test yourself regularly to check your progress.
- Practise unit test questions. This will highlight gaps in your knowledge and understanding, and will improve your exam technique.
- *Active* revision is much better than simply reading over material. Discussing topics, summarising notes and completing the various activities included in this book will all increase the effectiveness of your revision.

Activities toolkit

In the 'Testing your knowledge and understanding' sections, your answers to the shorter questions will show you how familiar you are with the main technical terms and concepts. The other questions in these sections contain activities that you can use to develop your own revision cards and materials. The activities help to improve your revision techniques and your learning, and by using them you can find out which ones work best for you.

There are different ways of presenting and summarising information and you need to find out which ones suit you. To help you with this there is a range of activities that you can try. These include:

- spider diagrams
- concept or mind maps
- annotated diagrams
- summary tables
- flow diagrams
- lists of vocabulary and definitions
- revision cards.

Spider diagram

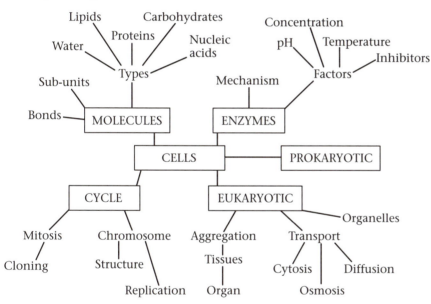

Spider diagrams and concept maps

Making up spider diagrams and concept or mind maps gives an overview of a topic or unit. The topics in each unit are a good starting point for making these. They should be structured in a clear and logical way. The diagram shown on the left is an example from Unit 1.

Annotated diagrams

Drawing diagrams of apparatus, objects or organisms should conform to a pattern that suits you. Labelling them by starting at the top right-hand side and drawing lines to the parts on the outside first and the centre last is one way of doing this. It is a good idea to add numbers or letters to the lines and to put a key below the diagram. Then you can cover up the key or the diagram and use it to test yourself. If the labels, along with short descriptions, are put on the diagram, then it will become an annotated diagram. The diagram of a respirometer on page 21 is an example of an annotated diagram.

Summary tables

Constructing tables that highlight similarities or differences helps you to make connections and summarise information. The activities in the units will give you examples of the features to select for a table. With practice you will be able to make the selection for yourself. The table on page xii is an example from Unit 3, *Energy and the environment*, Part a – Topic 1.

Mode of nutrition	Holozoic	Saprobiontic	Parasitic	Mutualistic
Example	human	*Rhizopus*	*Taenia*	Papilionaceae *Rhizobium*
Nutrient source	other organisms	dead, decaying organisms	host	carbohydrates from plant, nitrogen compounds from bacterium
Enzymes, digestion	own, hydrolytic extracellular	own, hydrolytic from hyphae, extracellular	from host, hydrolytic, extracellular	not applicable
Absorption surface	ileum (villi, microvilli, capillaries, lacteals)	hyphae, mycelium	integument, epithelium of segment	exchange occurs at the root nodule

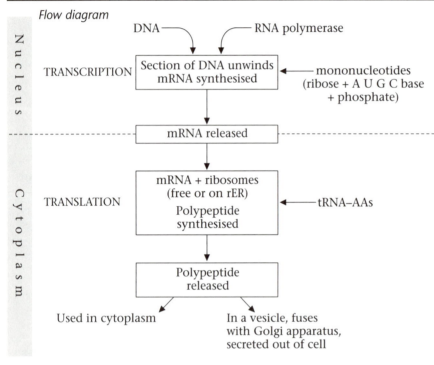

Flow diagram

Flow diagrams

Drawing flow diagrams to show the major steps and processes involved in a procedure helps your understanding because you may have to simplify your notes and present the information in a different way.

Revision cards

Some of the *Helpful hints* suggest making revision cards. Revision cards are easy to carry around and use when you have a little spare time, for example when waiting for that bus to arrive. A systematic approach to labelling these makes it easier to find the right card when you want to revise a particular unit or topic. Try to follow the same system that you have for your notes; put the unit title and the topic at the top and then number each card for the topic. The card could have any of the outcomes of a revision activity on it. It can also be used as a testing exercise if part of the card can be used as a question and the answers on the card are covered up. The example on the left illustrates this. It has a list of vocabulary on the left-hand side and the descriptions on the right. The cells are listed in order of lignification and size. The *Helpful hints* will usually suggest additions that could be made. In this case the cells found in phloem tissue could be added.

UNIT 2 (Exchange, transport and reproduction)
(b) Transport in flowering plants

Xylem (general) — supporting tissue, transport inorganic solutes, mainly lignified cells

Xylem fibres — single long thin cell, usually lignified walls, mainly support

Xylem tracheids — single cell, tapering ends, pitted lignified walls, mainly support

Xylem vessels — several cells as a tube, lignified walls, no end walls, no cell contents, long distance transport

Xylem parenchyma — single thin-walls, packed between and around vessels

Preparing for the unit tests

It is important that you are aware of the structure of the unit tests, including the number of papers, the time allowed for each paper, the types of questions used, any optional topics and the nature of the coursework or practical assessment (see Table 2, page viii).

Information about the content of the Edexcel unit tests is provided in the specification and is summarised for AS Biology and·Human Biology in Table 1 (page vi). It is useful to familiarise yourself with some past papers, or specimen papers, to get an idea of what to expect.

There are a number of terms commonly used in unit tests, and it is important that you understand the meaning of each of these terms.

Calculate	Carry out a mathematical calculation, showing your working and providing the appropriate units.
Compare	Point out similarities *and* differences.
Define	Give a statement outlining what is meant by a particular term.
Describe	Provide an accurate account of the main points (an explanation is not necessary).
Discuss	Describe and evaluate, putting forward the various opinions on a topic.
Distinguish between	Point out differences only.
Explain	Give reasons, with reference to biological facts and theories.
Outline	Provide the essential facts.
State / Give / Name	Give a concise, factual answer.
Suggest	Use biological knowledge to put forward an appropriate answer in an unfamiliar situation.
What? / Why? / Where?	Direct questions requiring concise answers.

Whatever the question style, you must read the question *very carefully*, underline key words or phrases, think about your response and allocate time according to the number of marks available. Further advice and guidance on answering test questions is provided throughout this book.

Structured questions

These are short-answer questions that may require a single word answer, a short sentence, or a response amounting to several sentences. Answers should be clear, concise and to the point. The marks allocated and the space provided for the answer usually give an indication of the amount of detail required. Typical question styles include:

- naming parts on diagrams
- filling in gaps in a prose passage
- completing tables and tick boxes
- performing calculations
- interpreting experimental data and graphs
- describing processes and the functions of structures.
- devising or planning practical investigations.

Free prose questions

Each unit test will have at least one question that requires you to produce a longer answer. These questions enable you to demonstrate the depth and breadth of your biological knowledge, as well as your ability to communicate scientific ideas in a concise and clear manner. The following points should help you to perform well when answering free prose questions.

- You should make your points clearly and concisely, illustrated by examples where appropriate.

- Try to avoid repetition and keep the answer relevant (refer back to the question).

- The points you make should cover the *full range* of the topic addressed in the question.

- Use diagrams only if appropriate and where they make a useful contribution to the quality of the answer.

- Spend the appropriate amount of time on the question (proportional to the marks available).

The day of the unit test

Try to feel confident, and then you will be able to do your best. Make sure you know which unit test you are sitting and the type of questions you can expect (see Table 2 on page viii). If you find it helpful, spend a little time looking over some Biology before you go into the test room in order to get your mind working along biological lines. Make sure that you get to the test room in plenty of time.

Check that you have:

- two or more blue or black pens, and two or more pencils

- your calculator – if the batteries are old, replace them

- a watch to check the time

- a ruler.

Don't take a red pen with you as the awarding body doesn't allow you to use this colour – the examiners use red for marking the papers.

Tackling the question paper

- Read each question very carefully so that your answers are appropriate.

- Make sure that you write legibly (you cannot be given marks if the examiner cannot read what you have written) and try to spell scientific terms accurately.

- Do *not* use correcting fluid. As well as wasting time while you wait for it to dry, you may also forget to go back and fill in the gap (and examiners can't give marks for something that's not written down!). Instead, neatly cross out what you have written. If, later, you realise that what you first wrote is correct, write 'ignore crossing out'. The examiner will then mark it.

- If you need more room for your answer, look for space at the bottom of the page, the end of the question or after the last question, or use supplementary sheets. If you use these spaces, or sheets, alert the examiner by adding 'continued below', or 'continued on page …'.

You need to think carefully about time or it may become a problem.

- Make sure that you know how long you have got for the whole test and how many questions you have to do in this time. Pace yourself and try to use the number of marks that are allocated to a question as a guide to how long you should spend on that question.

As a rough guide, you are allowed about one minute per mark available. So you should spend roughly 15 minutes on a question worth 15 marks.

- Do not write out the question, but try to make a number of valid points that correspond to the number of marks available. If you get stuck, make a note of the question number and move on. Later, if you have time, go back and try that difficult question again.

- It is a good idea to leave a few minutes at the end to check through the paper, correcting any mistakes or filling in any gaps.

Final note

We hope that you find *Make the Grade in AS Biology with Human Biology* useful and that it makes a contribution to your future success.

1 Molecules and Cells

Introduction

The content of this unit is common to both the Biology and the Biology (Human) specifications.

This unit of the specification includes the basic information on biological molecules that you will require for the whole course. You have probably met some of the topics in this unit before, either at Key Stage 4 or in other introductory courses, but here you will study them in more depth and detail. It is vital to ensure that you learn all the relevant details as outlined in the specification.

Within this unit, there is a progression. For example, a knowledge of **protein structure** helps you to understand the section on **enzymes**, and the section on nucleic acids leads to a consideration of the way in which **proteins are synthesised** within cells. A detailed knowledge of **cell structure** helps you to understand that several **organelles** are involved with the synthesis and export of proteins from cells. It is relevant at this stage to know where these organelles are located within cells and to appreciate their interrelationships.

Consideration is given to the differences between **prokaryotic** and **eukaryotic cells**, and you will need to be familiar with the structures of these cells, so that you can identify organelles from electronmicrographs. Some of the required practical work for this unit, involving the use of a microscope, also reinforces your knowledge of the structure of eukaryotic cells and the tissues which they form.

The ways in which molecules and ions move into and out of cells is another fundamental topic and it is important to understand the basic principles of **osmosis**, **diffusion**, **facilitated diffusion** and **active transport**. A sound understanding of these methods of transport will help you in all other units of the specification.

It should be stressed that this unit contains a great deal of the basic information on which many other units depend. If you can make sure that you have a good knowledge of all the topics, you will have gained a solid foundation for your further studies in this subject.

The unit is divided into four topics:
1 Biological molecules
2 Enzymes
3 Cellular organisation
4 The cell cycle

Topic ❶ Biological molecules

Introduction

The structure and roles of some of these molecules, together with an understanding of the way in which atoms can combine to form compounds by the formation of **chemical bonds**, are required at Key Stage 4. A much more detailed knowledge is expected for AS and Advanced GCE. A sound understanding of this part of the specification is fundamental to many of the later topics and you can save yourself a great deal of trouble by learning the material thoroughly at the beginning of the course.

It is necessary to be able to recognise and identify the **chemical structures** and **general formulae** of certain molecules and the best way of committing these to memory is to practise writing them out. In this way, you will become familiar with the numbers and arrangement of the atoms. Some knowledge of chemistry can be useful, but there is help available in the Physical Science Background at the end of Nelson Advanced Science Molecules and Cells, which covers the physical science background that you need for the course. Other information can be added, to build up your knowledge of the functions and locations of the molecules. It is better to concentrate on learning one group at a time, so that you do not become confused.

The practical work for this section involves using qualitative and quantitative tests for starch, reducing and non-reducing sugars and proteins. These are all very straightforward tests and are described on page 4. Do make sure that you know these thoroughly as you could be asked questions on them.

Checklist of things to know and understand

Before attempting to answer any questions, check that you know and understand the following:

Water

- ❏ the formation of hydrogen bonds

- ❏ its importance as a solvent

- ❏ the other roles related to its special properties, e.g. high latent heat of vaporisation, specific heat capacity, density and surface tension

Carbohydrates

- ❏ the structure and roles of the monosaccharides α- and β-glucose, ribose and deoxyribose

- ❏ the roles of fructose and galactose

- ❏ the monomers, structure and roles of the disaccharides maltose, sucrose and lactose

- ❏ the structure and roles of the polysaccharides starch, cellulose and glycogen

Lipids

☐ the general nature of lipids as fats, oils and waxes

☐ the general structure of a triglyceride

☐ the roles of lipids as energy stores

☐ the nature of saturated and unsaturated fatty acids

☐ the structure and properties of phospholipids and their role in the structure and properties of cell membranes

Proteins

☐ the general formula and general structure of amino acids

☐ the nature of amino acids as monomers in the formation of polypeptides and proteins

☐ the terms *primary*, *secondary*, *tertiary* and *quaternary structure* and their importance in the structure of enzymes

☐ the roles of ionic, hydrogen and disulphide bonds in the structure of proteins such as insulin and collagen

☐ the nature and roles of fibrous and globular proteins as illustrated by collagen and insulin

Nucleic acids

☐ the basic structure of a mononucleotide and that RNA and DNA are composed of mononucleotides involving the bases thymine, uracil and cytosine (pyrimidines) and adenine and guanine (purines)

☐ the structure of DNA, including base pairing and the nature of the double helix

☐ the mechanism of replication of DNA

☐ the structure and roles of messenger and transfer RNA

☐ the nature of the genetic code

☐ that a gene is a sequence of bases on the DNA coding for a sequence of amino acids in a polypeptide chain

☐ the processes of transcription and translation in the synthesis of proteins

☐ the function of the ribosomes in protein synthesis

☐ that amino acid sequences are specified by DNA

☐ codons and anticodons in relation to messenger and transfer RNA

☐ an appreciation of the Human Genome Project in the light of the structure and roles of nucleic acids, together with consideration of the spiritual, moral, ethical, social and cultural issues involved.

In addition, check that you understand the following principles, which may apply to more than one group of molecules:

☐ that condensation and hydrolysis reactions are involved in the synthesis and degradation of disaccharides and polysaccharides, polypeptides and proteins

☐ the nature of the bonds linking monomers in the formation of polymers, e.g. glycosidic, ester, peptide

❏ that condensation reactions are involved in the formation of mononucleotides and polynucleotides in RNA and DNA.

Practicals

You are expected to have carried out practical work to investigate:

❏ *biochemical tests for starch using iodine (qualitative and quantitative)*

❏ *biochemical tests for reducing and non-reducing sugars using Benedict's reagent (qualitative and quantitative)*

❏ *biochemical test for protein using biuret reagent.*

> ## Practical work – Helpful hints
>
> The practical work for this section involves using qualitative and quantitative tests for starch, reducing and non-reducing sugars and proteins. These are all very straightforward tests. Make sure that you know these thoroughly as you could be asked questions about them.
>
> ● Iodine solution is used as a test for starch, giving a blue-black colour when starch and iodine are mixed together.
>
> ● Benedict's reagent is used to test for reducing sugars. When they are mixed together and heated, a range of colours, including yellow, orange or brick-red, will be produced if reducing sugars are present.
>
> ● Biuret reagent is used as a test for proteins, giving a violet colour when mixed with protein.
>
> ● Sucrose is a non-reducing sugar, which can be hydrolysed into its constituent monosaccharides (glucose and fructose) using either dilute hydrochloric acid, or sucrase. Both glucose and fructose are reducing sugars, which can then be detected using Benedict's reagent.
>
> These tests can be made quantitative, using a colorimeter with starch and iodine, or with protein and biuret reagent, by using a range of standards and plotting a calibration curve. For reducing sugars you could prepare a range of colour standards, using glucose solutions of known concentrations. Alternatively, glucose test strips, such as Diabur 5000™, can be used.

Testing your knowledge and understanding

The answers to the numbered questions are on pages 86–103.

To test your knowledge and understanding of biological molecules, try answering the following questions.

Monosaccharides

1.1 Write down a general formula for a monosaccharide in terms of C, H and O atoms, using *n* to represent the numbers of each.

1.2 List all the monosaccharides you are familiar with, grouping them according to the number of carbon atoms they have.

1.3 Write down a general formula for each group giving the number of C, H and O atoms.

1.4 One of the groups is known as the hexoses. How many C atoms in a hexose?

1.5 Give the names of *three* hexoses.

1.6 Where would you expect to find each in the human body?

1.7 Suggest their roles in humans.

Disaccharides and polysaccharides

You could test your knowledge of disaccharides and polysaccharides in a similar way, remembering that both types of molecules are made up of monosaccharide monomers.

The following are good ways of learning the disaccharides.

1.8 Write a list of those you know.

1.9 Give a location where each can be found occurring naturally and give its role. Your answer could be in the form of a table.

1.10 State the constituent monomers of each disaccharide.

1.11 Work out the general formula in terms of C, H and O atoms.

1.12 Draw a structural diagram of the two monomers and indicate how condensation occurs to form a glycosidic bond.

Similar exercises can be carried out for **polysaccharides** and this approach can be modified to assist with learning the detailed information needed for **water**, **lipids**, **proteins** and **nucleic acids**. You can make up your own straightforward factual questions, or you can modify those given above so that they are relevant to the particular group of molecules you are revising.

Other activities that you might like to try to help you learn this topic are given below.

Polysaccharides

1.13 Construct a table of the properties of polysaccharides, to include their sub-units, location, roles in plants and animals, and solubility. This can be used for revision and for testing your knowledge by covering up the columns.

Triglycerides and phospholipids

1.14 Make a spider diagram containing as much information about the structure and roles of these molecules as you can. Remember to give reasons for specific roles, e.g. an explanation of why triglycerides are important for insulation or as energy stores.

All these activities are suitable for **revision cards**. It is often helpful to organise your knowledge of one topic in several different ways because it assists your ability to assess what is needed for an answer and enables you to respond more quickly without wasting time in the examination. Most topics are easily summarised in the ways shown above and a quick glance can be useful in testing your knowledge or reminding you of a detail.

Helpful hints

A spider diagram is extremely useful if you need to give an account of the structure and properties of a particular group of compounds, so you could do one for all the groups of biologically important molecules. See Activities toolkit, page xi.

Unit 1

Mark allocations are given for each part of the questions and the answers are given on pages 86–103.

Helpful hints

- Make your crosses and ticks very clear and unambiguous: if you change your mind, cross through your original answer and write your corrected answer to one side.
- If you find this type of question difficult, you could use the boxes to write in the correct information. For example, if the molecule does not have glycosidic bonds, then write in the type of bonds it does have.
- Try making up your own tables for different molecules in one group of compounds and then for different groups of compounds.

- There are three marks for the drawing, so you have to think carefully about what you need to show in order to gain the maximum mark.
- Some of the functions may relate to other topic areas, so you may need to think about what happens to polypeptides.

 Practice questions

1 The table below refers to triglycerides and glycogen. If the statement is correct for the molecule, place a tick (✔) in the appropriate box. If it is incorrect, place a cross (✗) in the appropriate box.

Statement	Triglyceride	Glycogen
contains only carbon, hydrogen and oxygen		
glycosidic bonds present		
soluble in water		
provides storage of energy		
occurs in flowering plants and animals		

(Total 5 marks)
(Edexcel B / HB1, June 1998, Q. 3)

2 Polypeptides are synthesised from amino acids. The diagram below shows the molecular structure of an amino acid.

$$
\begin{array}{ccccc}
H & & H & & O \\
| & & | & & \| \\
N & - & C & - & C \\
| & & | & & \\
H & & R & & OH \\
\end{array}
$$

(a) (i) Draw a molecular diagram to show how this amino acid reacts with another amino acid to produce a dipeptide. **[3]**

(ii) Name the type of reaction involved. **[1]**

(b) State *two* functions of the R groups in a polypeptide chain. **[2]**

(Total 6 marks)
(Edexcel B / HB1, January 1998, Q. 3)

3 The diagram below shows the structure of part of a molecule of deoxyribonucleic acid (DNA).

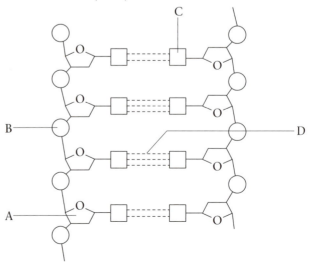

Helpful hints

● Be as specific as possible in naming the parts of the molecule. On an examination paper, you are given labelled lines on which to write your answers and you should check that you have not muddled the parts.

● Again think very carefully about your diagram and what you need to include in order to gain the maximum marks.

(a) Name the parts labelled A, B, C and D. **[4]**

(b) Draw a ring around one nucleotide. **[1]**

(c) (i) Draw a diagram to show how the process of replication takes place. **[3]**

   (ii) Explain why DNA replication is described as semi-conservative. **[2]**

(d) Name the stage in the cell cycle during which DNA replication occurs. **[1]**

(Total 11 marks)
(Modified from Edexcel B / HB1, June 1996, Q. 2)

4   Give an account of the structure and functions of polysaccharides in living organisms.

(Total 10 marks)
(Edexcel B / HB1, June 1996, Q. 8)

You could practise this type of question by substituting other compounds, such as proteins, triglycerides or nucleic acids, although some of these compounds could form the basis of longer answers or essays.

● This is known as a *free prose* question and you need to include factual information. The spider diagrams and tabulated information on your revision cards are excellent ways in which to prepare for such questions.

● In an examination you may gain most of your marks for *structure*, but remember to include some points about the *functions*.

● Try to make your answer as concise and precise as possible. You might find it helpful to jot down a few reminders, but there is no need to spend time making an elaborate plan as this is not an essay question.

Topic ② Enzymes

Introduction

Before you can understand how enzymes work you need to know about their structure. In order to do this you will have to refer to your notes about the types of **chemical bonds** and how these contribute to the structure of **proteins**. You need to remember that the primary, secondary and tertiary structure of a protein is related to the bonding within the protein. The residual groups (R groups) of amino acids can be charged, hydrophilic or hydrophobic and so contribute to the properties, structure and activity of the proteins of which they are a part. The R groups can interact with each other to form ionic, hydrogen or covalent bonds, which help to stabilise the protein structure. Other molecules and ions can interact with the R groups. This is the case when a change in hydrogen ion concentration causes a change in pH. The bonds between the R groups can be broken if the pH or temperature of the enzyme's environment is altered. These can cause a change in the shape and properties of the protein, a process known as *denaturation*.

Knowledge of enzymes and how they work will be required when you study the biochemical processes involved in digestion, synthesis of nucleic acids and proteins.

Unit 1

 Checklist of things to know and understand

Before attempting to answer any of the questions, check that you know and understand the following:

❑ that enzymes are proteins which act as catalysts

❑ that enzymes can change the activation energy of a reaction

❑ how the structure of an enzyme contributes to its specificity and catalytic activity

❑ the concept, nature and role of the active site

❑ the effect of changing enzyme and substrate concentrations on the rate of the reaction

❑ the effect of changing temperature and pH on the rate of the reaction

❑ how active site-directed and non-active site-directed inhibition of enzyme action occurs

❑ how immobilised enzymes can be used

❑ that enzymes are used commercially

❑ that pectinases are used in food modification

❑ that proteases are used in biological detergents

❑ the advantages of using immobilised enzymes

❑ that lactase can be used in an immobilised form.

Practicals

You are expected to have carried out practical work to investigate:

❑ *the effects of enzyme concentration, temperature and pH on enzyme-catalysed reactions*

❑ *enzyme immobilisation using lactase*

❑ *the use of pectinase in the production of fruit juice.*

Practical work – Helpful hints

The practical work for this section involves experiments involving enzymes, to investigate factors affecting enzyme activity and investigations using immobilised enzymes. There are a number of general points which you should be able to describe, which can be applied to investigations using different enzymes.

When giving an account of a practical to investigate enzyme activity remember the following points.

● Always refer to volumes, or concentrations of solutions rather than using vague terms such as 'amount' or 'quantity'.

● Temperature affects enzyme activity so it is important to keep temperature constant, using a water bath.

● Allow the enzyme and substrate solutions to reach the correct temperature before they are mixed together, in other words, the solutions are allowed to equilibrate to the temperature of the water bath.

● The rate of reaction may then be found by finding the reciprocal of the time taken to reach the end point, i.e. 1 ÷ time to reach the end point.

● Enzyme activity is also affected by changes in pH.

● The pH of a solution can be kept constant by using a buffer solution (see Appendix p107).

● If you are investigating the effect of pH on enzyme activity, then a range of buffer solutions can be used, for example, at pH values of 5.0, 6.0, 7.0 and 8.0.

● Immobilised enzymes are enzymes which are trapped in an inert material, such as calcium alginate. Briefly, the enzyme solution is mixed with sodium alginate, then allowed to drip into a solution of calcium chloride. Beads of insoluble calcium alginate form, containing the trapped enzyme. Immobilised enzymes are used in numerous industrial processes, such as in the preparation of high-fructose corn syrup, using immobilised glucose isomerase.

● Pectinase is an enzyme which can be used both to increase the yield of fruit juice from fruits such as apples and to clarify juice by breaking down pectins, which can make the juice appear cloudy.

 Testing your knowledge and understanding

The answers to the numbered questions are on pages 86–103.

Helpful hints

The terms and statements would make a useful revision card.

To test your knowledge and understanding of enzymes, try answering the following questions.

2.1 List *five* characteristics of enzymes.

2.2 Using E to represent the enzyme, S to represent the substrate and P to represent the product, write down an equation for the enzyme-controlled reaction.

2.3 List the factors that affect an enzyme-catalysed reaction.

2.4 There are several terms used when describing the activity of enzymes. Write down the meanings of the following terms: *active site*, *maximum rate of reaction*, *optimum pH*, and *turnover number*.

2.5 Each of the statements below describes a feature of enzyme-catalysed reactions. Give a word or phrase that matches each of the statements.

(a) Change in bonding changes the tertiary structure of the protein and the shape of its active site so that it is less complementary to the substrate and the substrate is less able to bind to the enzyme.

(b) A non-substrate molecule binds to the active site so preventing the entry of the substrate.

(c) A non-substrate molecule binds to another part of the enzyme, changing the shape of the enzyme and the active site.

2.6 Draw a sketch graph that shows how the rate of an enzyme-catalysed reaction is affected by a change in pH. Put on the graph an X to represent the maximum rate of reaction and a Y to represent the optimum pH. Now annotate the graph by explaining why there is an increase in rate, a peak and a decrease in rate.

2.7 Make a list of all the enzymes you need to know about. Now construct a table that includes the substrate, product and use or location of each enzyme.

Practice questions

1 The statements below refer to three enzymes. Complete the table by filling in the blank spaces.

Enzyme	Substrate	Type of reaction	Product	Commercial use
pectinase	pectin		monosaccharides disaccharides galacturonic acid	
lactase		hydrolysis		
DNA polymerase				DNA fingerprinting

(Total 8 marks)
(New question)

2 Explain the meaning of each of the following terms.

(a) Immobilised enzyme [2]
(b) Denaturation [2]
(c) Activation energy [2]

(Total 6 marks)
(New question)

Topic 3 Cellular organisation

Introduction

In order to understand the structure and activity of organelles and the relationships between them, you need to know about the molecules of which they are made up. You need to read your notes on **phospholipids**, **carbohydrates**, **nucleic acids** and **enzymes**. An understanding of the

---

see Activities toolkit, page xi.

Helpful hints

You could draw an annotated sketch graph for each of the factors that affect enzyme-catalysed reactions. Each graph should have labelled axes and show the general shape of the curve expected, as well as indicating which other factors are kept constant. The main parts of the curve should be labelled with a description and an explanation of the effect the factor has in that part.

Mark allocations are given for each part of the questions and the answers are given on pages 86–103.

Helpful hints

● Explanations should be as concise as possible.

● Sometimes appropriate examples can help with the explanation and may be asked for by the question.

nature of cellular membranes is fundamental to this section. You need to remember the structure of a phospholipid and that it has a hydrophilic (phosphate) region and a hydrophobic (fatty acid) region. This means that they will be organised so that the hydrophilic regions are in contact with the aqueous medium on either side of a membrane. Proteins may have hydrophobic regions and these will be in contact with the hydrophobic regions of phospholipids. Carbohydrates that are made up of a small number of monomers are called glycosides, because of the glycosidic bond between each monomer.

Knowledge and understanding of this section is particularly important when you study transport systems and osmoregulation.

 Checklist of things to know and understand

Before attempting to answer any of the questions, check that you know and understand the following:

❑ the limits of magnification and resolution of light and electron microscopes

❑ how to calculate the magnification of a drawing or photomicrograph

❑ the meaning of the terms *prokaryotic* and *eukaryotic*

❑ the roles of the parts of a bacterial cell: cell wall, flagella, mesosomes, chromosome, plasmids, glycogen granules and lipid droplets

❑ the structure of a bacterial cell

❑ the structure of a leaf palisade cell

❑ the structure of a liver cell

❑ the structure and composition of the cell surface (plasma) membrane

❑ the properties and roles of the cell surface (plasma) membrane

❑ the structure and composition of cell components

❑ the roles of the nucleus, endoplasmic reticulum, Golgi apparatus, lysosomes, chloroplasts, mitochondria, ribosomes, centrioles and microtubules, cellulose cell wall

❑ that molecules and ions move in and out of cells

❑ the principles of diffusion, facilitated diffusion, water potential, osmosis, active transport, endocytosis and exocytosis

❑ that tissues are aggregations of cells

❑ the structure and function of the tissues of a mesophytic leaf

❑ that organs are aggregations of tissues

❑ that a leaf and the liver are organs.

Practicals

You are expected to have carried out practical work to investigate:

☐ *the setting up and use of a light microscope to view slides of cells and tissues*

☐ *making accurate drawings of cells and plans of tissues*

☐ *using a graticule to determine the scale or magnification of the drawings.*

Practical work – Helpful hints

In this section, you are expected to be able to use a microscope to observe suitable cells and tissues and make accurate drawings. You are also expected to be able to use a graticule to measure cells. Questions may be set which ask you to describe the use of a microscope and a graticule.

When describing the use of a microscope, remember the following points.

● Always start off with the low-power objective, making sure that it is correctly in position.

● Switch on the light, or adjust the mirror to illuminate the field of view evenly.

● Place the slide on the stage so that the specimen is directly below the objective lens.

● Carefully bring the specimen into focus using the coarse focus knob.

● When the specimen is in focus with the low-power objective, carefully turn the next power objective into position. It should now only be necessary to use the fine focus knob to bring the specimen sharply into focus.

● After viewing the slide, always leave the microscope with the low-power objective lens in position.

● An eyepiece graticule is a disc (made of glass or plastic) with a scale on it. This is placed in the eyepiece lens of the microscope.

● The graticule is calibrated using a stage micrometer. This is a special type of microscope slide with an accurately ruled scale etched on it. The eyepiece scale is superimposed over the stage micrometer scale so that the eyepiece graticule may be calibrated.

● When you have made a drawing of cells using the microscope, you should always include a scale bar, to show the actual dimensions of the cells.

 ## Testing your knowledge and understanding

 The answers to the numbered questions are on pages 86 and 103.

To test your knowledge and understanding of cellular organisation try answering the following questions.

Organelles

3.1 Calculate the magnification of an organelle in a photomicrograph. The length of the organelle is 10.5 μm and the same length on the photomicrograph is 15 cm.

3.2 Organelles in eukaryotic organisms can be grouped depending on the type of membrane they have surrounding them.

(a) Draw a simple diagram of the structure of a cell surface (plasma) membrane.

(b) List all the organelles in a cell that have no membrane.

Helpful hints

Make sure that the units of length are the same or convert them to the same unit before you do the calculation. If your calculator gives answers in standard form make sure that the × 10 part is not missing from your answer. Magnification does not have a unit.

Helpful hints

You could annotate your diagram (see 3.2 (a) on p12) by giving brief descriptions of the roles of the parts of the membrane.

Helpful hints

You could extend your table to include all of the organelles or produce tables that show differences between similar organelles.

Helpful hints

You could make up your own questions about the other components of the cell surface (plasma) membrane or links between organelles.

 Mark allocations are given for each part of the question and the answers are given on pages 86–103.

(c) List all the organelles with a single membrane.

(d) List all the organelles with a double membrane (envelope).

3.3 A good way of testing your knowledge of organelles is to rank them according to size and to look for similarities and differences between them.

(a) List all the organelles that contain nucleic acids and state the type of nucleic acid.

(b) Draw out a table that shows the similarities between mitochondria and chloroplasts.

Transport mechanisms

Your understanding of the transport mechanisms operating in cells can be improved by first learning the basic principles of each of them.

3.4 Write down the main conditions required for diffusion across a membrane to occur.

3.5 Write down the main conditions for osmosis to occur.

3.6 Write down how facilitated diffusion differs from active transport.

Functions of the components of cell membranes and organs

3.7 Write down all the functions of glycosidic side chains.

3.8 What is the link between the nucleus, centrioles and microtubules?

3.9 List in order, the tissues present in a mesophytic leaf, starting with the upper epidermis and give a function for each one.

 Practice question

1 The table below refers to three organic compounds found in cell organelles. If the compound is found in the organelle, place a tick (✔) in the appropriate box and if the compound is not in the organelle, place a cross (✗) in the appropriate box.

Organelle	Phospholipid	DNA	RNA
ribosome			
chloroplast			
Golgi apparatus			
smooth endoplasmic reticulum			
mitochondrion			

(Total 5 marks)
(Modified from Edexcel B/HB1, June 1996, Q. 1)

Unit 1

Topic ❹ The cell cycle

Introduction

The **cell cycle** consists of the sequence of events from the formation of a new cell, its growth, division of the nucleus (**mitosis**) and subsequent division of the cytoplasm (**cytokinesis**) to form two new daughter cells. In order to understand this cycle, it is necessary to have a good background knowledge of cell structure and a sound grasp of the structure of DNA, together with an understanding of the way in which the replication of the DNA occurs.

Once you have an understanding of the structure of DNA, you need to appreciate the relationship between the DNA and the proteins in the structure of the chromosomes. You will need to know the details of all the events in the cell cycle, not just the stages in mitosis.

Checklist of things to know and understand

Before attempting to answer any of the questions, check that you know and understand the following:

- ❑ that a chromosome in the nucleus of a eukaryotic cell consists of DNA and histones

- ❑ how the replication of DNA occurs and the roles of the enzymes involved

- ❑ the sequence of events in the cell cycle

- ❑ the events of prophase, metaphase, anaphase and telophase

- ❑ the behaviour of the chromosomes during the stages of mitosis

- ❑ the significance of mitosis in growth and replacement in that daughter nuclei have identical numbers and types of chromosomes

- ❑ that the process of mitosis enables the transfer of identical genetic information from parent to offspring

- ❑ that a leaf palisade cell and a liver cell have a diploid chromosome number and have been produced by nuclear division followed by differentiation

- ❑ the nature of natural and artificial cloning in plants and animals.

Practicals

You are expected to have carried out practical work to investigate:

☐ *Preparation and staining of root tip squashes to recognise and study stages in mitosis using a light microscope.*

 Practical work – Helpful hints

Practical work in this section requires you to make a root tip squash to observe stages of mitosis. Garlic bulbs are often used for this practical, as they produce roots quite quickly when they are allowed to grow supported over water.

● The end of the root is used as this contains cells which are actively dividing.

● This can be stained using either acetic-orcein, or the Feulgen method, which will stain nuclei and chromosomes.

● The root tip is carefully squashed on a microscope slide to spread out the cells.

● Examine the squash preparation using low power first, then high power.

● You could try counting the number of cells that show stages of mitosis, and express this as a percentage of the total number of cells present in the field of view. This percentage is known as the *mitotic index*.

If you do not get good results from your own preparations of root tip squashes, there are plenty of excellent diagrams of the stages of mitosis in the text books.

  *Testing your knowledge and understanding*

 The answers to the numbered questions are on pages 86–103.

To test your knowledge and understanding of the cell cycle, try answering the following questions.

4.1 The diagram below shows the cell cycle, with its different stages and their durations.

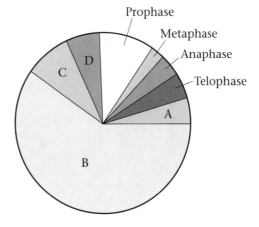

(a) In which stage will protein synthesis be highest and why?
(b) In which stage is new DNA synthesised?
(c) In which stage do cell organelles replicate?
(d) In which stage does cytokinesis occur?

(e) In which stage are new organelles synthesised?

(f) In which stage is an energy store built up and why?

4.2 Try the following questions on the replication of DNA.

(a) Prior to replication, how do the two strands of the double helix separate?

(b) Where does the replication take place?

(c) Explain why the replication is known as *semi-conservative*.

(d) State the function of DNA polymerase.

(e) What is the function of DNA ligase?

4.3 The following statements are all events in the process of mitosis. For each statement, decide which stage of mitosis it relates to and designate P (for prophase), M (for metaphase), A (for anaphase) or T (for telophase) accordingly. When you have four groups of statements, arrange them in chronological sequence, giving each statement a number.

(a) microtubules organised into spindle

(b) chromatids separate

(c) chromosomes shorten, thicken; spiralisation occurs

(d) chromosome seen to consist of two chromatids joined at the centromere

(e) nucleoli reappear

(f) chromosomes lengthen and uncoil

(g) chromosomes appear as long tangled threads

(h) nucleolus disappears

(i) chromatids (now daughter chromosomes) reach the poles of the cell

(j) centromeres divide

(k) chromosomes arranged on equator of spindle

(l) nuclear envelope forms round each group of daughter chromosomes

(m) spindle fibres shorten, pulling centromeres to opposite poles

(n) centrioles move to opposite poles of cell

(o) nuclear envelope disappears

(p) chromosomes attached by centromeres to spindle fibres

Helpful hints

You could try drawing diagrams of the events listed (a to p) on separate cards. When you have them in the right order, you could try flicking through them quickly to animate the process!

Mark allocations are given for each part of the questions and the answers are given on pages 86–103.

Practice questions

1 Read through the following passage on the cell cycle and mitosis, then provide the most appropriate word or words to complete the passage.

In the cell cycle, replication of the DNA takes place during .... At the beginning of prophase, the chromosomes become visible and can be seen to consist of two .... joined at the .... The .... and .... disappear and a spindle develops in the cell.

The chromosomes become attached to the spindle during .... At .... one copy of each chromosome is pulled towards each .... of the spindle. The final phase, called .... , involves the formation of two new nuclei. In plant cells, the two daughter cells are separated by the formation of a ....

(Total 10 marks)

(Modified from Edexcel B / HB1, January 1997, Q. 3)

2   The graph below shows how the quantity of DNA, measured in arbitrary units, varies with time during the different phases of the cell cycle in an animal cell.

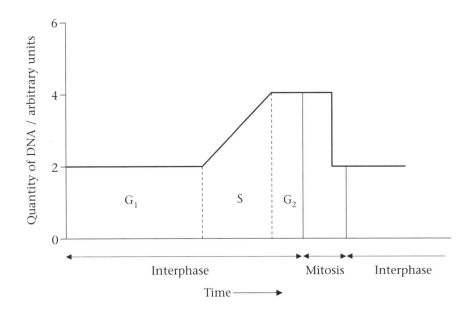

(a)   Explain what is happening within the cell during phase S. **[2]**
(b)   Explain what is happening during phase G_1. **[2]**
(c)   Name *three* processes that are occurring during G_2 which do not occur during G_1. **[3]**
(d)   Explain why the quantity of DNA in the cell changes during mitosis. **[2]**

(Total 9 marks)

(Modified from Edexcel B / HB1, January 1996, Q. 4)

Unit 1 Assessment questions

1   The statements in the table below refer to three polysaccharide molecules. Complete the table by filling in the blank spaces.

Mark allocations are given for each part of the questions and the answers are given on pages 104–106.

Helpful hints

Think carefully about starch – you might want to make the boxes bigger!

Property / Role	Starch	Glycogen	Cellulose
constituent monomers	α-glucose		
type of bonds present			1,4 glycosidic
nature of polysaccharide chain		very branched	
role in living organisms	energy store in plants		

(Total 8 marks)

(Modified from Edexcel B / HB1, January 1996, Q. 4)

2 The diagram below illustrates one model of enzyme action.

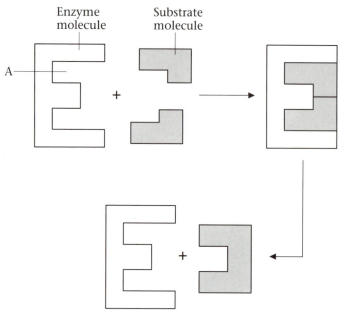

(a) Name the part of the enzyme labelled A. **[1]**

(b) Explain how this model can account for enzyme specificity. **[2]**

(c) With reference to this model, explain how inhibitors could affect this reaction. **[3]**

(Total 6 marks)

(Modified from Edexcel B / HB1, June 1998, Q. 2)

3 Read the following about the palisade cells of a leaf and provide the most appropriate word or words to complete the passage.

The palisade cell is typical of plant cells in that it has three structures: ...., ... and ...., none of which is present in animal cells. In common with animal cells, plant cells (such as palisade cells) have membrane-bound organelles which are not present in .... cells. In a leaf, palisade cells are grouped together as a layer just below the epidermis, forming a ...., the function of which is to carry out photosynthesis.

(Total 5 marks)

(Edexcel B / HB1, January 1998, Q. 2)

4 Explain what is meant by each of the following terms.

(a) Active transport **[3]**

(b) Enzyme immobilisation **[3]**

(Total 6 marks)

(New question)

5 Amyloglucosidase is an enzyme which breaks down starch to glucose by removing glucose units stepwise from starch molecules.

The effect of temperature on the activity of a commercial preparation of amyloglucosidase is shown in the graph below.

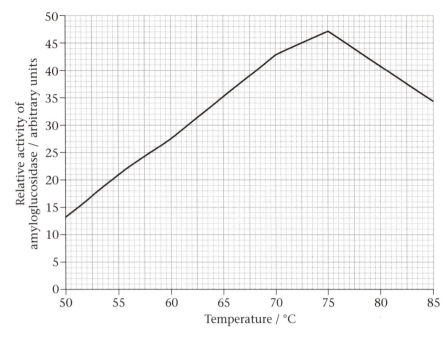

(a) (i) Calculate the percentage increase in activity of the enzyme as the temperature increases from 65 °C to 75 °C. Show your working. **[3]**

(ii) Explain why temperature affects enzyme activity as shown in the graph. **[4]**

(iii) This enzyme works at a relatively high optimum temperature. Suggest why this may be an advantage when using the enzyme in industrial processes. **[2]**

(b) Suggest what effect the addition of heavy metal ions, such as mercury or silver, would have on the activity of amyloglucosidase, giving a reason for your answer. **[2]**

(Total 11 marks)
(Edexcel B / HB1, January 2000, Q. 7)

6 Give an account of the structure and replication of deoxyribonucleic acid (DNA).

(Total 10 marks)
(Edexcel B / HB1, June 1999, Q. 8)

2B Exchange, transport and reproduction

Introduction

Unit 2B is the Biology pathway. If you are studying Biology (Human) go to page 44.

This unit is about how organisms obtain the materials they need from their environment. The external surface area to volume ratio of an organism usually decreases as the number of cells it is made up of increases. Materials such as gases and nutrients have to get into organisms through their exchange surface. These surfaces show adaptations which increase the efficiency of entry of the materials and are related to the environment in which the organism lives. In multicellular organisms, once the materials have been absorbed, the gases and nutrients may have to be transported to the cells where they are used. The transport systems of organisms and how they are adapted are studied in this section. Usually organisms that are successful in the environment survive to breed. The sexual reproductive systems of organisms have certain things in common: production of **haploid** gametes, transfer of **gametes**, **fertilisation** and the development of the **embryo**. From Unit 1, your knowledge of **carbohydrates**, **lipids**, **proteins**, **enzymes** and mechanisms of cellular transport is used in this unit.

The unit is divided into four topics:

1B **Exchanges with the environment**
2B **Transport systems**
3B **Adaptations to the environment**
4B **Sexual reproduction**

Topic 1B

Exchanges with the environment

Introduction

Try to remember that all **exchange surfaces** have some features in common. A *large* surface area enables more exchange to occur. A *thin* surface provides a short distance and therefore faster **diffusion**. The presence of water, on a moist surface, enables solutes to dissolve. If a **diffusion gradient** is maintained across the surface then diffusion will be faster.

Checklist of things to know and understand

Before attempting to answer any of the questions, check that you know and understand the following:

Gas exchange and breathing

☐ that the efficiency of exchange depends on surface area to volume ratio

- how the features of exchange surfaces and ventilation mechanisms aid passive and active transport
- how gas exchange is achieved in protozoa / protoctista and a mesophytic leaf
- the structure of a mesophytic leaf
- how the opening of stomata is brought about
- the structure of the thorax, alveoli and the function of surfactants
- the terms *vital capacity* and *tidal volume*
- how breathing is controlled by the respiratory centre in the brain

Diet and digestion

- the structure of the alimentary canal
- how mastication and peristalsis are brought about
- the histology of the ileum wall
- the sources and effects of the secretions involved in the digestion of carbohydrates.

Practicals

You are expected to have carried out practical work to investigate:

- *the use of a simple respirometer.*

Respirometer

Time taken to travel set distance = rate drop moves to left

Drop of liquid

Scale

Capillary tube

Organism – known mass, same genetic stock

Soda lime or potassium hydroxide – absorbs carbon dioxide

Water bath – controlled at set temperature or varied to investigate effect of temperature

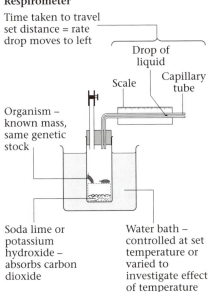

Practical work – Helpful hints

A simple respirometer is used to investigate the uptake of oxygen by respiring organisms, such as blowfly larvae, or germinating seeds. There are many different types of respirometers, but there are several general points which apply to their use. A simple respirometer is shown in the diagram (left). When describing the use of a respirometer, you should remember the following points.

- It is important to maintain a constant temperature during the experiment, so the respirometer should be placed in a water bath, set at a particular temperature, such as 25 °C.
- The respirometer usually contains potassium hydroxide, to absorb carbon dioxide. Take great care with potassium hydroxide as it is corrosive.
- Before taking any readings, the respirometer should be open to the air, e.g. by opening the screw clip to allow the organisms to adjust to the conditions and to allow the apparatus to equilibrate to the temperature of the water bath.
- When the clip is closed, the movement of the manometer fluid or the coloured drop in the capillary tube should be recorded at regular intervals.
- If you know the diameter of the capillary tubing, you can calculate the volume of oxygen taken up by the organisms.
- The rate of respiration is normally expressed in terms of the volume of oxygen taken up per unit time per unit mass of living material.
- You could use a respirometer to investigate the rate of anaerobic respiration of yeast cells, by measuring the volume of carbon dioxide produced. In this case, no potassium hydroxide is used.

Testing your knowledge and understanding

> The answers to the numbered questions are on pages 86–103.

Helpful hints

IB.1: This could be done as a simple diagram showing the outlines of the cells. You could add label boxes that list the features and roles of each of the types of cell.

Helpful hints

IB.4: You could even do this as a little story and when you have studied oxygen transport and respiration you could extend it to include these.

Helpful hints

You could add to your diagram the labelled connections to the respiratory centre in the brain.

To test your knowledge and understanding of exchanges with the environment, try answering the following questions.

Gas exchange and breathing

1B.1 List *five* features of a good gas exchange surface.

1B.2 Starting with the upper epidermis of a mesophytic leaf, write down the names of each of the tissue types present in the lamina.

1B.3 Describe the processes involved in stomatal opening.

1B.4 List the structures that a molecule of oxygen would pass through on its route from the external air to the blood plasma.

1B.5 How is inflation of the lungs brought about?

1B.6 Describe how a simple repirometer can be used to determine the rate of respiration of an organism.

1B.7 The diagram of the thorax below shows the main features involved in ventilation.

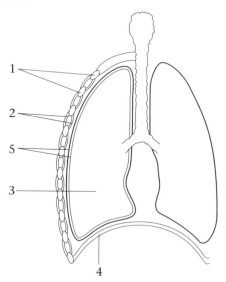

Sketch the diagram and below it draw numbered boxes. Inside each box name the structure and describe its role in ventilation.

Diet and digestion

1B.8 List in order all the parts of the alimentary canal.

1B.9 Name all the enzymes involved in the digestion of carbohydrates.

Helpful hints

You could do this in the form of a table that also shows the source of the enzyme and the products of hydrolysis.

Helpful hints

You could add the details of the rest of the ileum wall and annotate the diagram for one of your revision cards.

1B.10 Draw a simple diagram of a villus and label all the main parts.

You could produce a revision card that has a list of all the technical terms in this section such as *turgidity*, *vital capacity*, *tidal volume*, *mastication* and *peristalsis* and alongside them add descriptions of each one. See Activities toolkit, page xii.

Mark allocations are given for each part of the question and the answers are given on pages 86–103.

Practice question

1   The diagram below shows a section through lung tissue from a healthy person.

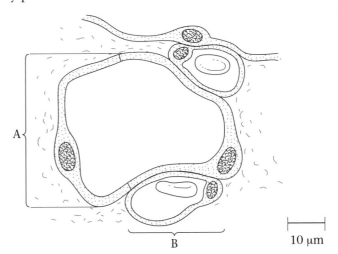

(a)   Name the parts labelled A and B.   [2]

(b)   Describe and explain how A and B are adapted for the normal functioning of the lungs.   [3]

(c)   Structure A secretes a surfactant. Explain how this substance helps the normal functioning of the lung.   [2]

(Total 7 marks)

(Modified from Edexcel HB3, June 1998, Q. 2)

Topic 2B Transport systems

Introduction

This section of the unit relates to the transport of materials in both flowering plants and mammals. You need to understand why transport systems are required and that such systems are related to the **size** and **surface area to volume ratio** of organisms. It is necessary to understand the concept of **mass flow** and to appreciate the movement of molecules within organisms. Some aspects of Unit 1 help in understanding the content of this section. For example, the nature of the carbohydrates glucose and sucrose, the properties of water and the functions of biological molecules, together with transport across membranes, should be reviewed. Some understanding of basic cell structure will also help in the sections on the movement of water and solutes in flowering plants.

This section can be divided into two distinct subtopics: **transport in flowering plants** and **transport in mammals**. The subtopic on flowering plants deals with the structure of the **vascular tissues** and the transport of water, mineral ions and the products of photosynthesis. In mammals, the transport involves the **blood** and the **circulatory system**.

Transport in flowering plants

You need to learn the composition of the transporting tissues, the phloem and the xylem, very thoroughly. Details of the different types of cells are important, especially as structure is very closely related to function in these cells. A clear understanding of the uptake and movement of water and mineral ions depends on a sound knowledge of the processes of diffusion, osmosis and active transport from Unit 1. A consideration of transpiration requires an understanding of the roles of stomata, so a knowledge of leaf structure is also advisable.

Much of this topic can be learnt through the use of annotated diagrams because structure and function are very closely linked. The structure of the flowering plant is not a favourite topic with students, but once you have got to grips with it and learnt the fundamentals, you will find the questions that are set are usually very straightforward to answer.

see Activities toolkit, page xi.

Checklist of things to know and understand

Before attempting to answer any of the questions, check that you know and understand the following:

❑ the structure of the vascular tissues

❑ that xylem is composed of vessels, tracheids, fibres and xylem parenchyma

- ❏ the role of vessels in relation to transport
- ❏ that phloem is composed of sieve tube elements, companion cells, phloem fibres and phloem parenchyma
- ❏ the role of sieve tube elements and companion cells in relation to transport
- ❏ the structure of a dicotyledonous root
- ❏ the uptake of water and its transport across the root to the xylem
- ❏ the way in which water is moved through the plant
- ❏ the apoplast, symplast and vacuolar pathways
- ❏ the role of the endodermis
- ❏ the functioning and roles of the transpiration stream
- ❏ the structure of vessels in relation to the cohesive and adhesive forces of water
- ❏ the contribution of vessels to the movement of water through the plant
- ❏ the roles of stomata
- ❏ the effect of different environmental conditions on the transpiration stream

Movement of nutrients

- ❏ the roles of diffusion and active transport in the uptake of mineral ions by roots
- ❏ the transport of mineral ions through the plant
- ❏ the translocation of organic solutes
- ❏ the difference between the transport of water and organic solutes
- ❏ the structure and arrangement of sieve tube elements, companion cells and transfer cells to the movement of organic solutes.

Practicals

You are expected to have carried out practical work to investigate:

- ❏ *demonstrations and measurements of transpiration rates using a potometer*
- ❏ *stomatal counts.*

Practical work – Helpful hints

In this section, you are expected to use a potometer to investigate transpiration and to make stomatal counts using epidermis from suitable leaves.

- A potometer is used to measure the volume of water taken up by a leafy shoot, which is almost the same as the volume of water lost in transpiration.

- The apparatus should be set up under water, and the stem cut under water, to avoid air bubbles.

- A bubble is introduced into the end of the capillary tube, and the movement of the bubble along the scale is carefully timed.

- Remember that factors such as light intensity and temperature will affect the transpiration rate and should therefore be controlled.

- The volume of water taken up can then be calculated and the rate of uptake found.

- You could work out the surface area of the leaves and express the results as volume of water taken up per unit time per unit leaf area.

- The effect of wind speed on the rate of transpiration could be investigated by using an electric fan to blow the leaves.

- Stomatal counts are carried out either by peeling the epidermis from a leaf and mounting on a microscope slide, or by preparing a nail varnish impression.

- The number of stomata present in the field of view can then be counted, and the mean of at least three different fields of view found. Find the diameter of the field of view using a stage micrometer, then calculate the number of stomata per unit area.

- You could use this method to make quantitative comparisons of the numbers of stomata present on leaves from different species of plants.

 Testing your knowledge and understanding

The answers to the numbered questions are on pages 86–103.

To test your knowledge and understanding of transport in flowering plants and movement of nutrients, try answering the following questions.

2B.1 Work out the surface area to volume ratio of:

    (a)   a cube with sides of 1 cm.

    (b)   a cube with sides of 2 cm.

    (c)   If you continue to increase the size of the cube, what happens to the surface area to volume ratio and what implications does this have for an organism?

2B.2 Name the cells which make up the xylem tissue.

2B.3 Name *two* types of cells found in phloem but not xylem.

2B.4 Give *four* environmental conditions which could affect the rate of transpiration. For each condition, state how the rate is affected.

2B.5 You can try out the following activity to test your knowledge of this topic.

Draw a large diagram of the cells in a root from the root hair to the xylem. Using coloured pens or pencils, draw in the **apoplast**, **symplast** and **vacuolar pathways**. You can also show the role of the *endodermis*.

Practice questions

Helpful hints

Remember that you are asked for *structural* features in the second column.

1 The table below refers to three different cells found in the vascular tissue of a flowering plant. Complete the table by providing the appropriate word or words for the empty boxes.

Cell type	One characteristic structural feature	One function
sieve tube element		
		transport of water and mineral ions
	narrow lumen	support

(Total 5 marks)
(Edexcel B3, January 1996, Q. 2)

Unit 2B

2 The graph below shows the results of an investigation into the relationship between the diameter of the stomatal aperture and the rate of transpiration in still and moving air. The stomatal transpiration was measured as water loss in nanograms (ng) second^{-1} per cm^2 of the leaf surface for stomatal aperture in the range 0 μm to 20 μm in diameter.

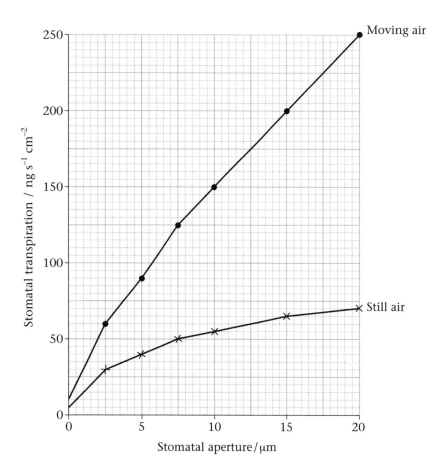

Helpful hints

When calculating the percentage change, read the figures from the graph carefully and as accurately as possible and remember which way round to do your calculation.

(a) Using data from the graph, calculate the percentage change in stomatal transpiration in moving air compared to still air when the stomatal aperture is 12.5 μm. Show your working. **[3]**

(b) Describe the relationship between stomatal transpiration and aperture diameter in still air. **[2]**

(c) Describe *two* ways in which the results for moving air differ from those for still air. **[2]**

(d) Suggest an explanation for the differences in stomatal transpiration in still air and in moving air. **[3]**

(e) Suggest *one* environmental factor, other than air movement, which would need to be kept constant in this experiment, giving an explanation for your answer. **[2]**

(Total 12 marks)
(Edexcel B6, January 1996, Q. 5)

Transport in mammals

Knowledge and understanding of diffusion, osmosis and active transport will help you to get a clearer picture of the relationship between plasma and tissue fluid. Knowledge of protein structure in Unit 1 will be useful when studying the transport of respiratory gases in erythrocytes. The blood is also important as a defence against disease.

 Checklist of things to know and understand

Before attempting to answer any of the questions, check that you know and understand the following:

- ❑ the structure of the mammalian heart, coronary circulation and the double circulatory system

- ❑ the cardiac cycle, myogenic stimulation and how the cardiac cycle is coordinated

- ❑ the structure and roles of arteries, capillaries and veins

- ❑ the composition of blood as plasma, erythrocytes and leucocytes

- ❑ how to study and identify blood cells (neutrophils, eosinophils, monocytes and lymphocytes)

- ❑ the structure of erythrocytes and their role in the transport of respiratory gases

- ❑ the transport of oxygen and carbon dioxide

- ❑ the roles of haemoglobin and fetal haemoglobin

- ❑ how to interpret dissociation curves of haemoglobin and the Bohr effect

- ❑ the interchanges between plasma and tissue fluid

- ❑ the roles of leucocytes in phagocytosis and in the secretion of antibodies.

Practicals

You are expected to have carried out practical work to investigate:

❏ *microscopic examination of stained blood films*

❏ *identification of blood cells.*

 Practical work – Helpful hints

In this section, you are expected to use a microscope to observe a stained blood film and to identify the different types of cells present.

● To see the blood cells clearly, it will be necessary to use an oil immersion lens.

● This is a high magnification objective lens (usually × 100) which is used with a small drop of immersion oil between the slide and the lens.

● Set up the microscope, place the slide on the stage and apply one drop of immersion oil to the slide directly beneath the oil immersion lens.

● Carefully rack down the oil immersion lens until it touches the oil, then focus carefully until the cells are brought sharply into focus.

● You should identify the types of blood cells present. You could carry out a differential white cell count, by noting the numbers of each type of white blood cell (leucocyte) which are seen. You may find it difficult to find any basophils, as these are the least numerous of all white cells.

● When you have finished using the oil immersion lens, always carefully wipe the oil from the slide and the lens, using a clean lens tissue.

 # Testing your knowledge and understanding

To test your knowledge and understanding of transport in mammals, try answering the following questions.

2B.6 List the functions of blood.

2B.7 Name all the numbered parts of the anterior view of the heart, shown on the diagram below.

The answers to the numbered questions are on pages 86–103.

 *Helpful hints*

You could use a diagram like this for a revision card and put the labels underneath so that you can use it to test yourself. You could add other information to the diagram such as the direction and relative pressure of blood and its composition.

Helpful hints

You could title this 'Double circulation', for one of your revision cards and explain underneath the advantages this system gives to a mammal.

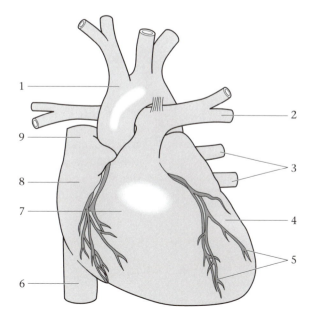

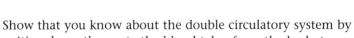
Helpful hints

There are some blood vessels that are exceptions to some of the general principles, e.g. the pulmonary vessels and hepatic portal vein.

Helpful hints

● Adding the functions of each of the components and diagrams of the cells showing distinguishing features, would give a more complete picture.

● You could produce a flow diagram of the stages in the cardiac cycle.

See Activities toolkit, pages xi–xii.

Helpful hints

A dissociation curve for fetal haemoglobin would show a curve to the left of one for adult haemoglobin. The fetal haemoglobin has a greater affinity for oxygen than maternal haemoglobin, when the blood has the same oxygen concentration.

 Mark allocations are given for each part of the question and the answers are given on pages 86–103.

2B.8 Show that you know about the double circulatory system by writing down the route the blood takes from the body to the aorta.

2B.9 List the ways in which veins differ from arteries.

2B.10 Relate the features of erythrocytes to their functions as efficient absorbers and transporters of respiratory gases.

2B.11 Most textbooks will have a summary table of the structures and roles of blood cells. Sometimes presenting information in a different way can improve your understanding of a topic.

Draw a spider diagram that shows the composition of the cellular components of the blood and their functions. Include all the types of leucocytes and put them on the diagram so that they are in a particular order. This could be based on size of the cell or the shape of the nucleus. An outline is given below to start you off.

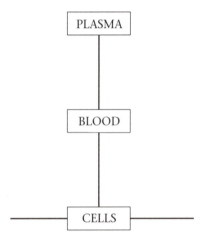

2B.12 In a dissociation curve for haemoglobin, an increase in carbon dioxide or decrease in pH shifts the curve to the right. This is known as the *Bohr effect*. Draw a dissociation curve for haemoglobin at two different concentrations of carbon dioxide. Underneath draw an outline to represent an erythrocyte and show how this effect is caused by the reactions inside the cell, when carbon dioxide is taken up.

2B.13 Using a series of diagrams show how antigens can cause the production of antibodies, and the antigens' destruction by phagocytes. Include as many of the types of leucocytes as possible and some of the details of phagocytosis.

Practice question

3 All the cells in the blood come from just one type of cell, the *multi-potential stem cell*. When the stem cell divides one of the two daughter cells may go on to give rise to other types of cell, whereas the other daughter cell remains a stem cell.

(*Source: The Triumph of the Embryo, Wolpert 1991*)

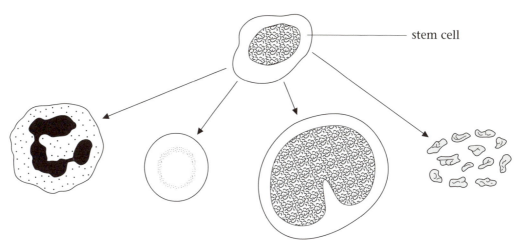

stem cell

(a) Suggest *one* region of the body where the stem cells referred to in the above extract are formed. **[1]**

(b) Name *two* types of blood cells that are phagocytic. **[2]**

(c) Which cellular component of blood begins the process of blood clotting? **[1]**

(d) Suggest why, when a stem cell divides, it is important that one daughter cell remains a stem cell. **[1]**

(Total 5 marks)

(Modified from Edexcel B3, June 1996, Q. 5)

Topic **3B** Adaptations to the environment

Introduction

In order to be successful an organism has to be well adapted to the environmental conditions present in its habitat. This section focuses on two environmental conditions and how organisms are adapted to these. The first is the availability of water to plants living in aquatic and terrestrial environments. Flowering plants that have xeromorphic features are adapted to habitats that have very little water available. The balance between the need for gas exchange, reducing water loss and maximising water uptake is brought about with the help of the adaptations shown by these organisms. Knowledge of the processes of **evaporation**, **diffusion**, **transpiration** and the factors that affect these will help you to understand how the adaptations work. The second environmental condition is the oxygen concentration in freshwater habitats. Invertebrates living in freshwater have a range of features that enable them to survive in an aquatic environment. Knowledge of the features of **gas exchange surfaces** and the **transport of respiratory gases** will help you to understand how these adaptations work. Reading about **indicator species**, found in clean and freshwater, will help you to become familiar with the range of adaptations shown by aquatic invertebrates. Knowledge and understanding of this section will help you when you study **human influences on the environment** in Unit 3.

Checklist of things to know and understand

Before attempting to answer any of the questions, check that you know and understand the following:

- ❑ the relationship of the external features of organisms to the physical characteristics of a specific habitat

- ❑ the xeromorphic features that enable flowering plants to live in habitats with low water availability

- ❑ the features of hydrophytes that enable them to live in aquatic habitats

- ❑ that freshwater can have varying oxygen concentrations

- ❑ the structural features of the gas exchange surfaces of freshwater invertebrates

- ❑ the physiological adaptations to varying oxygen concentrations shown by freshwater invertebrates.

Testing your knowledge and understanding

The answers to the numbered questions are on pages 86–103.

To test your knowledge and understanding of adaptations to the environment, try answering the following questions.

3B.1 Describe what is meant by the terms *xerophyte* and *hydrophyte*.

3B.2 List the environmental factors that increase water loss from the leaves of a plant.

3B.3 List the structures through which aquatic invertebrates obtain oxygen.

3B.4 Aquatic invertebrates obtain their oxygen either from that which is dissolved in the water or directly from the air and they will have structural and physiological features related to this. Their need for oxygen will depend on their level of activity and will also influence what sort of freshwater they will be found in. Water that flows over rocks or weirs, for example, will be aerated and have a high oxygen concentration but water polluted with sewage will have a low oxygen concentration.

The organism shown on the following page does not exist but it has all the features that aquatic invertebrates could use to ensure that they obtain sufficient oxygen. For each of the labelled features, state the name of an organism that uses this feature, where the oxygen is being obtained from and describe how the feature enables the efficient uptake of oxygen.

Helpful hints

You could convert this into a table or sketch the organism and add the information as annotations.

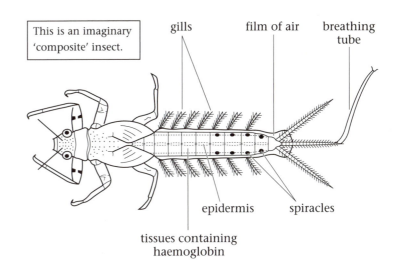

This is an imaginary 'composite' insect.

gills   film of air   breathing tube

epidermis   spiracles

tissues containing haemoglobin

Practice question

Mark allocations are given for each part of the question and the answers are given on pages 86–103.

1   The diagram below shows the larval stage of an aquatic invertebrate.

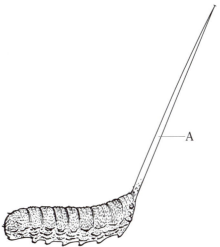

A

(a)   Name structure A.   **[1]**

(b)   Describe how structure A enables this organism to live in freshwater with a low concentration of oxygen.   **[3]**

(c)   Explain *one* other way in which this organism might obtain oxygen.   **[2]**

(Total 6 marks)
(New question)

Topic **4B**   Sexual reproduction

Introduction

This topic can be conveniently divided into three separate parts. The first part deals with **gamete formation** and the involvement of a reduction division in which the diploid chromosome number is reduced to haploid. It is advisable to have an understanding of the **cell cycle** and the process of **mitosis** from Unit 1 because many stages in gamete formation involve this type of cell division. You need to learn thoroughly all the events of both stages of **meiosis**, particularly with respect to **chiasmata** formation in prophase I. The events of meiosis show some similarity with those of mitosis, but it is important that you do not muddle the two processes. If you had a clear understanding of mitosis in Unit 1, then you will have no difficulty with meiosis here. Note that although you should know the names of the different stages in both divisions of meiosis, you do not need to know the names of the stages that make up prophase I. In most questions on this part of the topic, the events are the important things to learn; it is just convenient to be able to put them into a sequence under the headings 'Prophase', 'Metaphase', 'Anaphase' and 'Telophase'.

The second part of the topic concerns **reproduction in flowering plants**. You are required to know the structure of two flowers: an **insect-pollinated** dicotyledonous flower and a **wind-pollinated** grass. It is a good idea to choose flowers with which you are familiar or which are easily available for you to study. There is no substitute for observations made from living specimens, using a hand lens to study the different parts. Although there are excellent diagrams in most textbooks, it is very easy to dissect out the parts of living specimens and a better understanding of the relationship of the parts and the possible **pollination mechanisms** can be achieved. Insect-pollinated flowers such as wallflower, white deadnettle, primrose, buttercup, foxglove, *Antirrhinum* and sweet pea are suitable examples and usually readily available and the sizes of the internal structures can be seen easily with a hand lens or dissecting microscope. Grasses are more difficult and it is not always easy to find them in flower when you need them. The flowering parts are often quite small, but observations with a hand lens can reveal the anthers and the stigmas. The names of the parts and their functions should be learnt. Once the differences between the two types of flowers are appreciated, it is much easier to know and understand their adaptations to their mode of pollination and also to recognise the mechanisms which ensure that **cross-pollination** occurs.

The third part of the topic relates to **human reproduction** and also requires a knowledge of the structure and functions of the male and female reproductive systems. Too often, candidates do not give this part of the topic sufficient time, resulting in the inability to name parts correctly or to assign the correct functions to structures. The consequence of this is a loss of valuable marks in the examination. You should also have a clear understanding of **oogenesis** and **spermatogenesis**, together with an appreciation of the ways in which they differ. Detailed knowledge of the histology of the testis and ovary is not required, but you do need to

understand the roles of the female **hormones** involved in the **menstrual cycle**, during **pregnancy**, **birth** and **lactation**.

 Checklist of things to know and understand

Before attempting to answer any of the questions, check that you know and understand the following:

Gamete formation and meiosis

- ☐ that offspring result from the fusion of gametes producing a zygote

- ☐ that fusion of gametes leads to genetic variation in offspring

- ☐ that gamete formation involves a reduction division (meiosis) and its significance in the reduction of the diploid number of chromosomes to the haploid

- ☐ the behaviour of the chromosomes during both stages of meiosis, including chiasmata formation

- ☐ that haploid and diploid phases occur in the life cycle of organisms

Reproduction in flowering plants

- ☐ the structure and functions of the principal parts of an insect-pollinated dicotyledonous flower and a wind-pollinated grass

- ☐ how pollination occurs and the events leading to fertilisation

- ☐ the adaptations related to insect and wind pollination

- ☐ the significance of the mechanisms for ensuring cross-pollination

- ☐ protandry, protogyny and dioecious plants

Reproduction in humans

- ☐ the structure and functions of the human male and female reproductive systems

- ☐ the production of gametes in oogenesis and spermatogenesis

- ☐ the events of the menstrual cycle and the roles of follicle stimulating hormone, luteinising hormone, oestrogen and progesterone

- ☐ how male gametes are transferred leading to fertilisation

- ☐ how implantation occurs and the functions of the placenta in relation to the development of the embryo

- ☐ birth, lactation and the roles of oxytocin and prolactin.

Practicals

You are expected to have carried out practical work to investigate:

- ☐ *experimental investigation into the factors affecting the growth of pollen grains*

- ☐ *observations on preparations of insect testis squash for stages in meiosis.*

Helpful hints

An investigation into the factors which affect the germination of pollen grains would be very suitable for inclusion as part of the coursework as it could enable you to fulfil the criteria for planning.

Helpful hints

To prepare a hanging drop preparation to investigate the germination of pollen grains:

- Place a ring of plasticine on a microscope slide, so that the coverslip is supported about 4 mm above the surface of the slide.

- Place one drop of a sucrose solution in the centre of a coverslip and dust a small amount of pollen onto the sucrose solution.

- Invert the coverslip onto the ring of plasticine, so forming a *hanging drop preparation*.

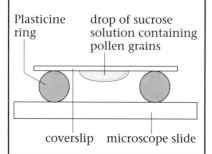

Plasticine ring   drop of sucrose solution containing pollen grains

coverslip   microscope slide

 The answers to the numbered questions are on pages 86–103.

Practical work – Helpful hints

This section requires you to investigate the growth of pollen grains and to observe a squash preparation of locust testes to see stages of meiosis.

Growth of pollen grains may be investigated by making a hanging drop preparation.

- Pollen grains are placed in a suitable growth medium, containing sucrose and boric acid (trioxoboric (III) acid).

- The preparations are examined using a microscope after 1 to 2 hours.

- The growth rate of pollen tubes could be investigated using a calibrated eyepiece graticule and measuring the length of the pollen tubes every 30 minutes for, say, 3 hours.

- You could investigate the effect of different concentrations of boric acid on the germination and growth rate of pollen tubes.

- Locust testes are dissected from a freshly killed male locust. Each testis consists of a number of finger-like follicles.

- Two or three follicles are placed on a microscope slide in a drop of saline solution and gently squashed with a scalpel blade.

- Acetic-orcein can be used to stain the nuclei and chromosomes, which can then be observed using a microscope.

- As an alternative, slides of permanent preparations of locust testes could be used.

 ## Testing your knowledge and understanding

To test your knowledge and understanding of sexual reproduction, try answering the following questions.

Gamete formation and meiosis

4B.1 Complete the following passage concerning the production of gametes by filling in the missing words.

Spermatogonia found lining the .... of the testis divide by .... to produce .... These cells then undergo .... eventually producing haploid cells called .... This division takes place in two stages and the intermediate cells are called .... A process of maturation, called spermiogenesis, occurs during which development into .... takes place.

During the formation of mature ova, primordial germ cells called .... , which have the .... number of chromosomes, become enlarged and are then known as .... Prior to ovulation, the first stage of a .... division occurs, resulting in the production of a .... and a .... The second phase of the division does not take place until .... occurs.

Reproduction in flowering plants

4B.2 Decide which of the features listed below is characteristic of insect-pollinated flowers (I) and which is characteristic of wind-pollinated flowers (W).

(a) Large petals

(b) Anthers hanging out of the flower

(c) Scent

(d) Fixed knob-like stigma

(e) Masses of smooth light pollen

(f) Coloured petals

(g) Feathery stigmas

(h) Flowers produced before leaves

(i) Presence of nectaries

(j) Lack of petals

(k) Anthers fixed to filaments

(l) Pollen with sculptured exine.

You could extend this exercise by stating how each of these features is associated with the method of pollination.

4B.3 Try identifying the following:

(a) The term given to poplar, willow and holly, which have male and female flowers on separate plants, making self-pollination impossible

(b) The tissue which is derived from the fusion of a male nucleus with the fusion nucleus (formed when the two polar nuclei fuse)

(c) The term used for flowers in which the stigma is receptive to pollen before the anthers release mature pollen

(d) The term used for the transfer of pollen from the anthers of one flower to the stigma of a flower on another plant of the same species

(e) The term used for flowers in which the stamens mature, releasing pollen from the anthers before the carpels mature.

Reproduction in humans

4B.4 The diagrams below show the human male and female reproductive systems.

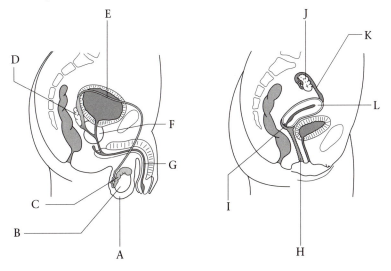

(a) Name the parts labelled A to L.

(b) Give the letter of the part which fits each of the following descriptions:

(i) Provides passage for the baby to leave the mother's body at birth

(ii) Produces the male gametes

(iii) Secretes oestrogen and progesterone

(iv) Secretes seminal fluid

(v) Collects and stores sperm

(vi) Produces ova

(vii) Provides passage through which spermatozoa are expelled

(viii) Conveys ovum to uterus.

4B.5 Your understanding of the hormones associated with the female reproductive system can be tested by constructing a table. You could head the columns 'Name of hormone', 'Source of secretion', 'Stimulus for secretion' and 'Effect'. You will need to leave plenty of room for the last column, as some of these hormones have more than one effect. Such a table could provide a valuable asset to your revision as it summarises all you need to know about the hormones and you can test yourself in different ways by covering up some of the columns.

Practice questions

Mark allocations are given for each part of the questions and the answers are given on pages 86–103.

1 The photomicrograph below shows plant cells which have undergone division during the formation of pollen grains.

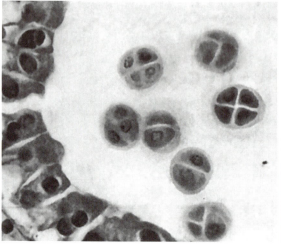

Magnification × 40

(a) Name the type of division shown in the photomicrograph. **[1]**

(b) **(i)** Name *one* location in a flowering plant in which this type of division occurs. **[1]**

(ii) Name *one* location in mammals in which this type of division occurs. **[1]**

(c) Give *two* reasons why this type of cell division is important in living organisms. **[2]**

(Total 5 marks)

(Edexcel B / HB1, June 1997, Q. 3)

2   Read through the following account of the hormonal control of the menstrual cycle and then provide the most appropriate word or words to complete the account.

> The release of .... from the anterior pituitary gland induces the development of .... Another hormone from the anterior pituitary gland causes the thecal cells to produce .... which controls the repair of .... after menstruation. At ovulation, a .... is released from the mature follicle. The remaining follicular cells form the .... which begins to secrete .... , inhibiting the release of the hormones from the anterior pituitary gland.

(Total 7 marks)
(Modified from Edexcel B3, June 1996, Q. 3)

Unit 2B Assessment questions

Mark allocations are given for each part of the questions and the answers are given on pages 104–106.

1   Read through the following account of stomata in flowering plants, then provide the most appropriate word or words to complete the account.

> Gases diffuse into and out of leaves through stomata. A stoma is a pore surrounded by two specialised cells known as .... cells. Changes in the turgidity of these two cells cause the diameter of the stomatal pore to vary. When .... ions enter these cells, their water potential becomes .... so water enters by osmosis. The resulting increase in turgor acts on their unequally thickened .... , causing the cells to change shape and the stomatal pore to ....

(Total 5 marks)
(Modified from Edexcel B3, January 2000, Q. 3)

2   An experiment was set up to investigate the uptake of different mineral ions by barley plants. A large number of barley seedlings were grown in a nutrient solution containing a range of mineral ions including potassium (K^+), calcium (Ca^{2+}), magnesium (Mg^{2+}) and nitrate (NO_3^-).
    The experiment was set up as shown in the diagram below.

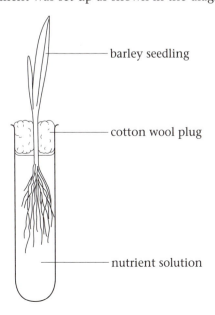

The concentrations of these ions in the solution were measured at the beginning and at the end of the experiment.
The results are shown in the table below.

Concentration of mineral ions / arbitrary units	At start of experiment	At end of experiment
nitrate	7	1.8
potassium	3	0
magnesium	2	2.1
calcium	5	5.6

(a) What do the results suggest about the mechanism of absorption of potassium ions? Explain your answer. **[3]**

(b) Suggest an explanation for the changes in concentrations of magnesium and calcium ions during the experiment. **[2]**

(c) Give *five* precautions which should be taken to ensure the results for all the barley seedlings were comparable. **[5]**

(d) Describe the pathway taken by mineral ions as they pass from the nutrient solution to the xylem in the roots of the seedlings. **[3]**

(Total 13 marks)
(Modified from Edexcel B3, January 1997, Q. 7)

3 The diagram below shows a transverse section of a leaf of *Ammophila arenaria*, which is a xerophyte. The photomicrograph shows the details of the area indicated by the box on the diagram.

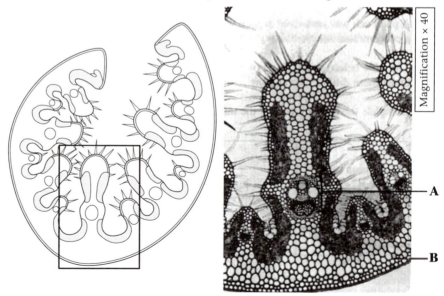

Magnification × 40

A

B

(a) Name the parts labelled A and B. **[2]**

(b) Describe *two* xeromorphic features shown in this leaf and, in each case, indicate how the feature helps to reduce transpiration. **[4]**

(Total 6 marks)
(Edexcel B2, January 1997, Q. 2)

4   The diagram below shows a vertical section through a heart. Blood vessels are not shown.

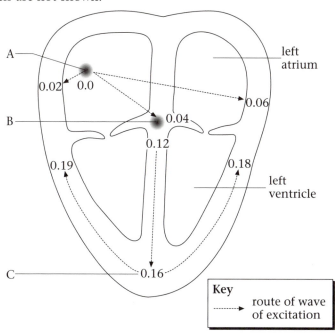

A and B are involved in the coordination of the heart beat. Waves of electrical excitation originate in A and spread throughout the heart. The figures on the diagram show the time taken in seconds for a wave of electrical excitation to reach different parts of the heart.

(a)   Name the areas labelled A and B.   **[2]**

(b)   Calculate the time taken for a wave of excitation to pass from B to C.   **[1]**

(c)   Using the information on the diagram, explain how the route taken by the wave of excitation contributes to the coordination of the heart beat.   **[4]**

(Total 7 marks)
(Edexcel B3, January 2000, Q.5)

5   The graph shows the oxygen dissociation curve for the pigment haemoglobin in a human. The loading tension is the partial pressure of oxygen at which 95 % of the pigment is saturated with oxygen. The unloading tension is the partial pressure at which 50 % of the pigment is saturated with oxygen.

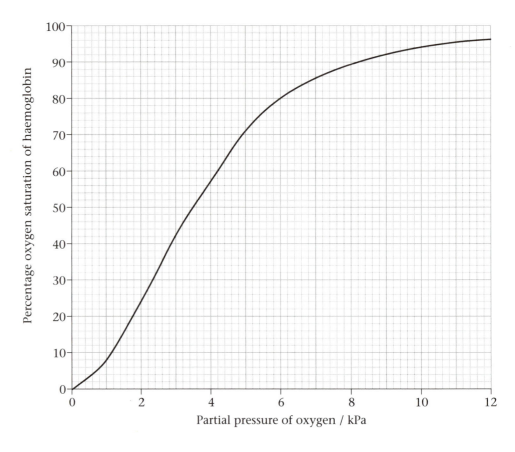

(a) Explain why haemoglobin is an efficient respiratory pigment. **[2]**

(b) (i)   From the graph, determine the difference between the loading and unloading tensions of haemoglobin. Show your working. **[2]**

  (ii)   Give *one* location in the body where partial pressures lower than the unloading tension may be reached. Give a reason for your answer. **[2]**

(c)   Suggest what effects increasing concentrations of carbon dioxide in the blood would have on the loading and unloading tensions of human haemoglobin. Give reasons for your answers. **[4]**

(d)   The oxygen dissociation curve for fetal haemoglobin lies to the left of the curve for adult haemoglobin. Suggest an explanation for this difference. **[2]**

(e)   State *three* ways in which carbon dioxide is transported in the blood. **[3]**

(Total 15 marks)
(Edexcel B6, June 1996, Q. 4)

6   The table below refers to the first and second divisions of meiosis. If the statement is correct, place a tick (✔) in the appropriate box and if the statement is incorrect, place a cross (✗) in the appropriate box.

Statement	First division of meiosis	Second division of meiosis
pairing of homologous chromosomes occurs		
chromosomes consist of pairs of chromatids during prophase		
chiasmata are formed		
chromatids are separated		
independent assortment of chromosomes occurs		

(Total 5 marks)
(Edexcel B / HB1, January 1998, Q. 1)

7   Give an account of the adaptations of flowers to insect and wind pollination as illustrated by an insect-pollinated dicotyledonous flower and a grass.

(Total 10 marks)
(Modified from Edexcel B3, January 1996, Q. 5)

2H   Exchange, transport and reproduction in humans

Introduction

Unit 2H is for the Biology (Human) pathway. If you are studying Biology go to page 20.

Some of the content of this Unit 2 Biology (Human) is the same as Unit 2 Biology, so you will be referred back to the relevant questions in Unit 2B. The external surface area to volume ratio of humans is relatively small and because of this the surfaces that exchange materials with the environment are specialised. To enable the exchange to occur efficiently these surfaces have specialised **epithelia** covering them. Once the materials have been absorbed, the gases and nutrients are transported in the **circulatory system** to the tissues where they are going to be used. The structure and functions of the heart, blood vessels and the blood, which make up the human circulatory system, are studied in this unit. Humans are able to adapt to very harsh environments such as those at high altitudes or regions with extremes of temperature. These **adaptations** can be short-term physiological changes that occur in visitors to the area but people who are native to such environments may show characteristics that have developed over generations. Humans who survive to breed may pass on their characteristics to their children. Sexual reproduction involves the production of **haploid** gametes, transfer of **gametes**, **fertilisation**, development of the **embryo** and **birth**. From Unit 1, knowledge of **carbohydrates**, **lipids**, **proteins**, **enzymes** and the mechanisms of cellular transport, is used in this unit.

The unit is divided into four topics:

1H   Exchanges with the environment
2H   Transport of materials
3H   Human ecology
4H   Human reproduction and development

Topic 1H   Exchanges with the environment

Introduction

Try to remember that all exchange surfaces have some features in common. A *large* surface area enables more exchange to occur. A *thin* surface provides a short distance and therefore faster diffusion. The presence of water, on a moist surface, enables solutes to dissolve. If a diffusion gradient is maintained across the surface then diffusion will be faster.

Checklist of things to know and understand

Before attempting to answer any of the questions, check that you know and understand the following:

Gas exchange and breathing

☐ respiratory gases, nutrients and excretory products are materials exchanged with the environment

☐ the features of exchange surfaces which aid passive and active transport

☐ the structure of squamous epithelium in an alveolus, cuboidal epithelium in a nephron and columnar epithelium in the ileum

☐ the structure of the breathing system and the mechanism of ventilation

☐ the principle of a spirometer and the interpretation of spirometer data

☐ the effects of physical activity and increase in carbon dioxide concentration on the breathing rate and tidal volume

☐ the characteristics of alveoli as surfaces involved in gas exchange

☐ the effects of smoking on ventilation and gas exchange

☐ the effects of smoking in relation to pregnancy

☐ the origin of carbon monoxide and its effects on gas exchange

Digestion and absorption

☐ the structure of the alimentary canal in relation to digestion and absorption

☐ mastication and the movement of food along the gut

☐ the histology of the ileum wall

☐ the sources and effects of secretions concerned with the digestion of carbohydrates.

Practicals

You are expected to have carried out practical work to investigate:

☐ *the use of simple apparatus to estimate vital capacity*

☐ *the effect of physical activity on breathing rates*

☐ *quantitative comparisons of the composition of inspired and expired air.*

Practical work – Helpful hints

In this section, you are expected to carry out simple experiments to measure vital capacity, breathing rates, and comparisons of inspired and expired air.

● A spirometer can be used to measure a person's vital capacity. Alternatively a hand-held spirometer, lung volume bags, or a calibrated bell jar, filled with water, may be used.

● Vital capacity is the maximum volume of air which can be exhaled, following a maximum inhalation.

● You could investigate the relationship between body weight and vital capacity, or differences between males and females.

● Breathing rate is difficult to measure, as watching the person can influence the rate.

● Investigate the effects of standardised exercise on breathing rate, and find out how long it takes for breathing rate to return to the resting value after exercise.

● Quantitative comparisons of inspired and expired air are made using either a J-tube or a gas burette.

● Potassium hydroxide solution is used to absorb carbon dioxide and alkaline pyrogallol to absorb oxygen.

● The changes in volume of gas in the J-tube or gas burette are used to calculate the proportions of oxygen and carbon dioxide in the air sample.

Testing your knowledge and understanding

The answers to the numbered questions are on pages 86–103.

To test your knowledge and understanding of exchanges with the environment, try answering the following questions.

Gas exchange and breathing

Try answering questions 1B.1, 1B.4, 1B.5 and 1B.7 from Unit 2B, page 22, and the following questions.

1H.1 Name *three* different types of epithelium and state where in the body they can be found.

1H.2 The diagram below shows a tracing from a spirometer. Sketch the diagram and explain what each of the labelled parts shows.

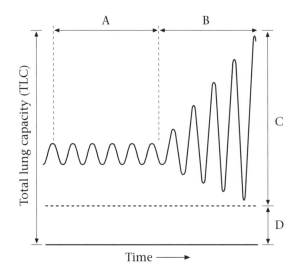

1H.3 Draw a table to summarise the effects smoking has on ventilation, gas exchange and pregnancy. Start with a list of the contents of tobacco smoke.

Digestion and absorption

Try answering questions 1B.8, 1B.9 and 1B.10 from Unit 2B, pages 22 and 23.

Practice question

Attempt practice question 1 parts (a) and (b) at the end of Topic 1B in Unit 2B, page 23 and part (c) below:

(c) (i) What effect might heavy smoking have on structure A?   **[1]**
    (ii) What result would this have on gaseous exchange?   **[1]**

Topic 2H Transport of materials

Introduction

Because humans are multicellular animals some of the cells are a long way from the surfaces through which materials are exchanged. This means that these materials have to be transported within the body. You already know about the gas exchange surface of the alveoli and the absorption of digested food in the villi. This part of the unit is concerned with how solutes are transported in blood, to cells, for use in metabolic processes such as respiration. The blood is also important as a defence against disease. Knowledge and understanding of **diffusion**, **osmosis** and **active transport** will help you to get a clearer picture of the relationship between plasma, tissue fluid and lymph. Knowledge of **protein structure** in Unit 1 will be useful when studying the transport of respiratory gases in **erythrocytes**.

Checklist of things to know and understand

Before attempting to answer any of the questions, check that you know and understand the following:

☐ the circulatory system transports respiratory gases, metabolites, metabolic waste and hormones

☐ the structure of the human heart and coronary circulation

☐ the cardiac cycle, myogenic stimulation and how the cardiac cycle is coordinated

☐ how to interpret a normal electrocardiogram (ECG)

☐ the role of artificial pacemakers

☐ the structure and roles of arteries, capillaries and veins

☐ the composition of blood as plasma, erythrocytes and leucocytes (neutrophils, eosinophils, monocytes and lymphocytes)

☐ how to study and identify blood cells

Unit 2H

❑ the structure of erythrocytes and their role in the transport of respiratory gases

❑ the transport of oxygen and carbon dioxide

❑ the roles of haemoglobin, fetal haemoglobin and myoglobin

❑ how to interpret dissociation curves of haemoglobin and the Bohr effect

❑ the interchanges between plasma, tissue fluid and lymph

❑ the roles of leucocytes in phagocytosis and in the secretion of antibodies.

Practicals

You are expected to have carried out practical work to investigate:

❑ *microscopic examination of stained blood films*

❑ *identification of blood cells*

❑ *the effects of physical activity on pulse rate.*

⊔⊔ *Practical work – Helpful hints*

In this section, you are expected to use a microscope to observe a stained blood film and to identify the different types of cells present.

● To see the blood cells clearly, it will be necessary to use an oil immersion lens.

● This is a high magnification objective lens (usually × 100) which is used with a small drop of immersion oil between the slide and the lens.

● Set up the microscope, place the slide on the stage and apply one drop of immersion oil to the slide directly beneath the oil immersion lens.

● Carefully rack down the oil immersion lens until it touches the oil, then focus carefully until the cells are brought sharply into focus.

● You should identify the types of blood cells present. You could carry out a differential white cell count, by noting the numbers of each type of white blood cell (leucocyte) which are seen. You may find it difficult to find any basophils, as these are the least numerous of all white cells.

When you have finished using the oil immersion lens, always carefully wipe the oil from the slide and the lens, using a clean lens tissue.

You are also expected to carry out a simple investigation into the effect of exercise on pulse rate. There are a number of electronic pulse meters which are available, such as 'wristwatch' style pulse monitors, or infra-red sensors which clip to the ear lobe. These can be useful to find pulse rates during activities, such as exercising. These monitors are, however, relatively expensive and there is nothing wrong with using the fingers on an artery method!

● Record the pulse rate of your subject at rest, by placing your index finger and second finger over a pulse point, such as the radial artery in the wrist.

● Record the pulse rate at regular intervals after your subject has performed some standardised exercise, such as cycling at a constant speed for two minutes on an exercise bicycle.

● You could compare the changes in pulse rates, in response to exercise, between two groups of people, such as those who take regular exercise and those who do not.

Testing your knowledge and understanding

The answers to the numbered questions are on pages 86–103.

To test your knowledge and understanding of transport in humans, try answering questions 1B.1, 1B.2, 1B.4, 1B.5, 1B.6, 1B.7 and 1B.8 from Unit 2B, page 22, as well as the following questions.

2H.1 Describe the role of artificial pacemakers.

2H.2 The diagram below shows part of an electrocardiogram. Describe what is happening in the heart to produce the region from P to T on the diagram.

Helpful hints

You could sketch this diagram, label all of the regions and add annotated descriptions to each label. See Activities toolkit, page xi.

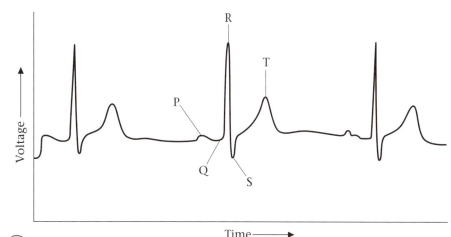

Practice question

Mark allocations are given for each part of the question and the answers are given on pages 86–103.

As well as attempting practice question 3 (page 30), in Unit 2B, Topic 2B, try the one below.

1 (a) State *two* features of capillaries that enable tissue fluid to be formed. **[2]**

(b) It is estimated that 85 % of tissue fluid is reabsorbed at the venous end of a capillary. Describe what happens to the remaining tissue fluid. **[3]**

(c) When tissue fluid is not sufficiently reabsorbed, it collects in the tissues causing swelling (oedema). Suggest one factor that could result in oedema. **[1]**

(Total 6 marks)

(Edexcel HB3, January 2000, Q. 4)

Unit 2H

Topic **3H** Human ecology

Introduction

In this topic you are looking at how the human body copes with living in extreme environmental conditions. Think about parts of the world where people cannot or do not live on a permanent basis. Then think why these regions are inhospitable to human life. You will probably realise that these regions are mainly those that are extremely cold or at high altitude (or both) and also extremely hot and dry (arid). As you go through the questions, try to understand why these conditions are at the limit of human tolerance and how they put the body under stress – which may lead to permanent damage to parts of the body or even death.

You should link your revision in this section to other topics in your AS course. You need to understand about **blood circulation** and the transport of respiratory gases (Unit 2H, Topics 1H and 2H). You also need to be familiar with the role of **haemoglobin** and its **dissociation curves**, **gas exchange** in humans including **ventilation** and the role of the **alveoli** (also in Unit 2H, Topics 1H and 2H).

Extremes of temperature

Checklist of things to know and understand

Before attempting to answer any of the questions, check that you know and understand the following:

❑ the normal body temperature and how far the body tolerates temperature variations

❑ the structures in the skin and their role in temperature regulation where relevant

❑ how temperature regulation is achieved through structural, physiological and behavioural mechanisms

❑ the causes and effects of heat stress and of cold stress

❑ how far natives and visitors show differences in their adaptation to extreme temperature conditions.

Testing your knowledge and understanding

The answers to the numbered questions are on pages 86–103.

To test your knowledge and understanding of extremes of temperature, try answering the following questions.

3H.1 'Normal' body temperature fluctuates around 36.8 °C. What term is used to describe temperature variation during a 24-hour period? At what time of day (or night) is body temperature at its lowest? About how much (in °C) is this daily variation? How low can core body temperatures fall before the situation becomes serious? When overheated, what is the upper limit to core body temperature at which death rapidly ensues?

3H.2 Look at the diagram of the mountaineer on the snow. Loss of heat from his body occurs in different ways, represented by the numbered arrows.

*Source: Barry PW (1996), Mountaineering, in Biological Sciences Review **8** (4) 23, Philip Allan Publishers.*

The list below gives some heat transfer mechanisms or ways that heat is lost from the body.

Select the correct mechanism from the list and put this into the highlighted box labelled A. In box B provide a suitable word or words to complete the statements which describe the heat transfer mechanism. In box C give features of clothing that would help to minimise losses of heat for the mountaineer in this situation.

List of heat transfer or heat loss mechanisms: *conduction; convection; evaporation of sweat; radiation;*

(a)

| **B** movement of air .... , which means that .... air next to the skin is replaced by .... air from outside |

A (arrow 1) ....

| **C** |

(b)

| **B** transfer of heat by .... with a solid surface |

A (arrow 2) ....

| **C** |

(c)

B can be an important mechanism for heat .... , e.g. from sun and .... from snow; also heat .... direct from body
A (arrow 3) ....
C

Helpful hints

Make sure you do not confuse the effects of heat loss and water loss, even though the two are interrelated at this point.

(d)

B .... of sweat from skin surface uses heat (latent heat of ....) results in ....
A (arrow 4) ....
C

3H.3 In hot climates, it is usually important to protect the body from *gaining* too much heat and to enable the body's cooling mechanisms to function adequately.

(a) Refer to the heat transfer and heat loss mechanisms listed in Question 3H.2 and describe how clothing and buildings in *hot dry* climates can incorporate features which help to minimise the effects of high temperatures. Focus your attention on features of traditional buildings and clothing worn by native residents at these locations rather than visitors.

(b) How might the features be different in a hot *humid* climate?

3H.4 In severely cold conditions, the body may suffer from cold stress. Damage to tissues may be in terms of **cold injury**, **frostbite** or **trench foot**. The whole body may be affected by **hypothermia**. Note that the sequence as given shows how the conditions become progressively more severe.

List these cold stress conditions and for each, describe the symptoms and the underlying physiological events in the body which lead to these symptoms or show how the damage is caused. You could make this into a useful summary table.

see Activities toolkit, pages xi–xii.

3H.5 In excessively hot conditions, the body may suffer from heat stress. There may be superficial damage to tissues, as in **sunburn**, and **prickly heat**. There may, however, be situations in which the whole body is affected, described as **heat collapse**, **heat exhaustion** and **heat stroke**. Note that the sequence as given again shows how the conditions become progressively more severe.

List these heat stress conditions and for each, describe the symptoms and the underlying physiological events in the body which lead to these symptoms. Again, you could summarise this in a table, like the one above for cold stress.

3H.6 Questions 3H.4 and 3H.5 looked at stresses the body may encounter in extreme conditions of external temperature variations. Heat exchange with the external environment has been summarised in question 3H.2.

(a) How does the body *gain* heat from the internal as well as the external environment?

(b) How may this heat gain change if the body undertakes exercise?

Even in less extreme conditions, the body experiences temperature fluctuations and its core temperature is kept within a narrow temperature range by **thermoregulatory mechanisms**.

Unit 2H

The skin is an important organ in controlling body temperature and maintaining the balance between heat gain and heat loss.

Helpful hints

In your annotations, label each of the relevant structures and indicate how it functions in both hot and cold conditions. See Activities toolkit, page xi.

(c) Where in the body are the *receptors* found which detect changes in the external environment?

(d) Which part of the body *detects* fluctuations in the temperature of the blood?

(e) Which part of the body acts as a 'thermostat' and triggers the *responses* to temperature fluctuations?

3H.7 Draw an annotated diagram of the skin so that you summarise features linked with temperature regulation. Your diagram can be quite simple, but design it so that it shows essential points, related to control in both hot and cold conditions. On your diagram, make sure you include reference to the following structures (but you may find more that are relevant): *temperature receptor; sweat gland, sweat duct; hair, hair erector muscle; adipose tissue; arterioles, blood capillaries, capillary loop.*

3H.8 To some extent, visitors to locations with extremes of temperature experience discomfort compared with natives living in these locations.

(a) List any features shown by such natives that may be seen as adaptations to hot climates and to cold climates.

(b) How far do visitors acclimatise to temperature extremes which are different from their normal place of residence? Think about both physiological and behavioural responses.

3H.9 Check your definitions: *thermogenesis* and *thermoregulation; vasoconstriction* and *vasodilation.*

High altitude

 Checklist of things to know and understand

Before attempting to answer any of the questions, check that you know and understand the following:

- ❏ the environmental conditions at high altitude that place stress on the body

- ❏ how the body's responses to lower oxygen availability lead to the symptoms of mountain sickness

- ❏ the symptoms of mountain sickness

- ❏ how the body systems respond to lower oxygen availability at high altitude

- ❏ how far natives and visitors show differences in their adaptation to high altitude.

Unit 2H

Testing your knowledge and understanding

The answers to the numbered questions are on pages 86–103.

Living at high altitude, whether as a native or as a lowland visitor, puts certain stresses on the body. These stresses arise from the environmental conditions associated with high altitude and become extreme at very high altitudes.

Unit 2H

To test your knowledge and understanding of high altitude, try answering the following questions.

3H.10 Assume that native high-altitude people are those who live permanently above 3000 m.

Name *two* areas in the world where people live permanently at high altitude.

3H.11 The list below gives *four* environmental conditions associated with high altitude.

For each of these, write down ways in which these conditions place stresses on the body. Think carefully because you may be able to find more than one way in which the condition affects the body.

You could organise your information in a table to make a useful summary. (The effects of low atmospheric pressure are considered in questions 3H.13, 3H.14 and 3H.15.)

List of environmental conditions: *low temperature; low humidity* (think of two effects); *high winds* (two effects); *increased solar radiation* (two effects);

3H.12 Refer to the conditions listed in Question 3H.11, and describe how clothing (or buildings) can incorporate features which help overcome these effects. Remember to include clothing worn by native highlanders, as well as visitors including mountaineers in high altitude regions.

3H.13 The group of charts which follow refer to some of the physiological effects of **hypoxia** (low levels of oxygen) and how this affects different systems of the body. The charts also indicate how the body may respond and other possible effects arising from the initial response. Some parts of the chart are incomplete but when you have completed them, you will have a useful framework which can act as a summary of the ways the body responds to hypoxia.

First select the correct 'response' from the list below and put this into the highlighted box (box A).

Then work through boxes B, C and D and find a suitable word (or words) to write on the dotted lines, to complete the statement.

The boxes represent the following: box A = response; box B = breakdown of response or how it is achieved; box C = effects in the body; box D = other effects where relevant (perhaps conflicting). [The first responses have been done for you.]

List of responses: *hyperventilation; increased haemoglobin; increased pulmonary diffusing capacity; decreased ADH secretion; changes in oxygen–haemoglobin dissociation; increased cardiac output; increased ADH secretion;*

(a) *Respiratory system*

(i)

	B *increased* rate of breathing (*more* breaths per minute) *deeper* breaths (*increased* volume per breath)
A *hyperventilation*	
	C .... intake of oxygen increases pO_2 at boundary of alveoli
	D BUT .... output of carbon dioxide leads to .... in blood (high pH) – leads to dizziness and nausea.

(ii)

	B .... surface area of lungs .... in blood flow at lung surface
A	
	C .... diffusion of oxygen from .... into blood in capillaries

(b) *Blood system*

(i)

	B .... heart rate (number of beats per minute) .... stroke volume
A	
	C .... blood pumped through pulmonary capillaries .... collection of oxygen at alveoli and transport to cells of body

(ii)

	B .... number of red blood cells (per unit volume) .... in haemoglobin concentration (mass per unit volume)
A	
	C .... carrying capacity for oxygen (BUT pO_2 must be high enough to allow Hb in the blood to collect oxygen)
	D BUT – blood becomes more viscous, which may .... blood flow in capillaries

Unit 2H

(iii)

A ....

B   dissociation curve shifts to ....

C   oxygen released .... in the cells
BUT some loss in % saturation of Hb with oxygen

(c) *Distribution of body fluids*
(i) *Mild hypoxia*

A ....

B   .... urine produced (diuresis)

(ii) *Severe hypoxia*

A ....

B   .... urine produced

C   .... fluid accumulates outside blood vessels,
particularly in lungs and brain. This condition is called
....
In .... oedema, the person may froth at the mouth
and become breathless. In .... oedema, fluid makes
the brain swell so it presses against the cranium and
causes headaches.

A visitor to high altitude can, to some extent, become acclimatised, though survival without supplementary oxygen at altitudes above 8000 m (such as close to the peak of Mount Everest) is limited to a few hours or less.

Another comment for you to consider – 'climbing the mountain is one thing, but getting down safely is the real test'! You might think about the physiology behind this remark.

3H.14 For each of the features referred to in question 3H.13, draw up a list to show how far native highlanders show adaptations which enable them to live successfully at high altitude.

3H.15 Acclimatisation can take from minutes through to days or weeks. Arrange the following 'effects' in sequence to give a 'time-scale' for acclimatisation, with the shortest time first: *increased red blood cell production; increased heart rate; increased capillary density; increased concentration of Hb in the blood; increased breathing rate.*

3H.16 (a) List some symptoms of mountain sickness and, as far as possible, relate them to physiological functions in the body, that are affected by high altitude.

(b) Generally, what is the best advice to mountaineers at high altitudes when suffering from acute mountain sickness?

Practice question

Mark allocations are given for each part of the question and the answers are given on pages 86–103.

1   An investigation was carried out into the effect of exposure to low environmental temperatures on core body temperatures. Two groups of males were studied: Europeans from a temperate climate and aborigines from a climate with hot days and cold nights. Both groups were exposed to an air temperature of 5 °C during a night of 8 hours. The results are shown in the graph on the following page.

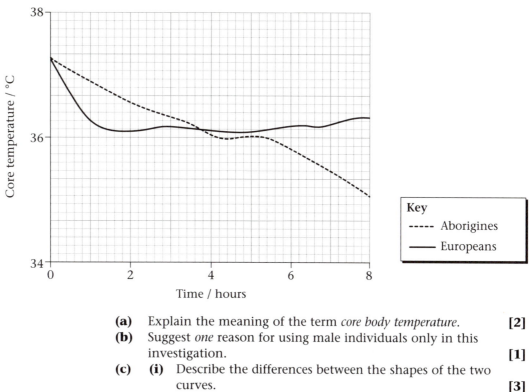

(a) Explain the meaning of the term *core body temperature*. **[2]**

(b) Suggest *one* reason for using male individuals only in this investigation. **[1]**

(c) (i) Describe the differences between the shapes of the two curves. **[3]**

   (ii) Suggest how these differences may be related to the climate in which the aborigines normally live. **[2]**

(d) Between 4 and 5 hours after exposure the mean core body temperatures remain constant. Explain how this is brought about. **[2]**

(e) Describe the effect on the body if the core temperature drops below 28 °C. **[3]**

(Total 13 marks)

(Edexcel HB2, January 1999, Q. 6)

Topic 4H Human reproduction and development

Introduction

If you are studying Human Biology, the specification is slightly different in that you do not need to learn about reproduction in flowering plants. The specification is set out differently, but is similar in that you do need the same knowledge of gamete formation and the structure and functions of the human reproductive systems. It is advisable to have an understanding of the **cell cycle** and the process of **mitosis** from Unit 1, because many of the stages in gamete formation involve this type of cell division. You need to learn thoroughly all the events of both stages of **meiosis**, particularly with respect to **chiasmata** formation in Prophase 1.

The events of meiosis show some similarity with those of mitosis, but it is important that you do not muddle the two processes. Note that you do not need to know the names of the stages that make up Prophase 1.

Knowledge of the structure and functions of the male and female reproductive systems is required. You should have a clear understanding of **oogenesis** and **spermatogenesis**, together with an appreciation of the ways in which they differ. Detailed knowledge of the histology of the testis and ovary is not required, but you need to know the sources and roles of the hormones involved in the menstrual cycle, and during pregnancy, birth and lactation. The section on human development following birth requires you to know about colostrum and milk production, how the proportions of the body changes and the effects of ageing.

 Checklist of things to know and understand

Before attempting to answer any of the questions, check that you know and understand the following:

❏ the structure and functions of the human male and female reproductive systems

❏ that gamete formation involves a reduction division (meiosis) and its significance in the reduction of the diploid number of chromosomes to the haploid

❏ the behaviour of chromosomes during both stages of meiosis, including chiasmata formation

❏ the production of gametes in oogenesis and spermatogenesis

❏ the events of the menstrual cycle and the roles of follicle stimulating hormone, luteinising hormone, oestrogen and progesterone

❏ how male gametes are transferred leading to fertilisation

❏ how implantation occurs and the functions of the placenta in relation to the development of the embryo

❏ the role of the placenta in controlling the passage of potentially harmful substances

❏ the stages of birth and the control of birth by fetal and maternal hormones

❏ colostrum and milk production

❏ the control of milk production by prolactin and oxytocin

❏ the importance of colostrum

❏ human growth curves, including changes in proportions of body parts from birth to maturity

❏ how ageing affects the skeletal, cardiovascular and reproductive systems.

Practicals

You are expected to have carried out practical work to investigate:

❏ *microscopic examination of the histology of ovary and testis.*

Practical work – Helpful hints

This section requires you to observe sections of ovary and testis in order to understand their structure and functions. Use prepared microscope slides showing sections of mammalian ovary and testis.

- Make low-power plans to show the outline structure of an ovary and a testis.
- Annotate your drawings to indicate the functions of the structures you have labelled.
- Make an annotated high-power drawing of a section through a seminiferous tubule.
- Remember to include scales on your drawings to show the actual size of the specimens.
- You may find it helpful to prepare flow charts to revise the processes of oogenesis and spermatogenesis. Indicate on your flow charts where mitosis, meiosis I and meiosis II occur.

 Testing your knowledge and understanding

To test your knowledge and understanding of human reproduction and development, you should check through the questions and activities suggested for the Biology specification. The ones which are suitable for you are 4B.1, 4B.4 and 4B.5. In addition, the following practice question relates specifically to your specification.

 Practice question

 Mark allocations are given for each part of the question and the answers are given on pages 86–103.

1 The hormone human chorionic gonadotrophin (HCG) is released from embryonic tissues during pregnancy. The effect of HCG is to prevent the degeneration of the corpus luteum in the mother's ovary.

The table below shows changes in the concentration in the blood of HCG and progesterone during the first 36 weeks of pregnancy.

Time / weeks	Concentration of human chorionic gonadotrophin (HCG) / arbitrary units	Concentration of progesterone / arbitrary units
0	0	7
2	0	7
4	15	8
8	60	9
12	45	10
16	24	11
20	12	13
24	10	15
28	10	20
32	14	30
36	12	55

Unit 2H

(a) Plot both sets of data in a suitable form on a piece of graph paper. [5]

(b) Describe and suggest explanations for the changes in HCG concentration during the first 20 weeks of pregnancy. [4]

(c) Describe *three* functions of progesterone. [3]

(d) Explain how the changes in progesterone level are related to its function. [2]

(Total 14 marks)

(New question)

Unit 2H Assessment questions

> Mark allocations are given for each part of the questions and the answers are given on pages 104–106.

1 Describe the effects of smoking on each of the following:

(a) Ventilation and gas exchange [3]

(b) Pregnancy [3]

(Total 6 marks)

(New question)

2 At high altitudes, atmospheric pressure is lower than at sea level. This means that people living at high altitude have less oxygen available to them compared with those living at low altitude. The table below shows the results of three measurements made on blood of normal individuals resident at each of three different altitudes.

Altitude / m	Percentage saturation of arterial blood with oxygen	Oxygen content of arterial blood / cm^3 per 100 cm^3	Haemoglobin in blood / g per 100 cm^3
150	94.2	18.3	15.5
3700	88.2	19.9	17.4
4375	83.5	20.9	18.6

(a) (i) How does the percentage saturation of arterial blood with oxygen change with increasing altitude? [1]

    (ii) Using the information in the table, explain how people living at high altitude have adapted to the low oxygen availability. [3]

    (iii) State *one* feature, other than those shown in the table, which enables high altitude residents to overcome the problem of low availability of oxygen. [1]

(b) State *three* environmental factors, other than the availability of oxygen, which vary between sea level and high altitude sites at the same latitude. [3]

(c) A visitor to high altitude from sea level may suffer from mountain sickness.

State *three* symptoms of mountain sickness. [3]

(Total 11 marks)

(Edexcel HB2, June 1998, Q. 6)

Now try assessment question 5 at the end of Unit 2B, page 42.

3 Energy and the environment

Introduction

The content of this unit is common to both the Biology and the Biology (Human) specifications.

There are two main components to this unit: Part a, which is assessed by means of a written test, and Part b, which is the coursework component, in which you have to present an individual investigation, supervised by your teacher. The main themes of Part a are **nutrition**, the source and **flow of energy through ecosystems**, the **effects on the environment of human activities** and some aspects of the **management of ecosystems**.

All living organisms depend on a supply of energy and heterotrophic organisms obtain their energy in the form of complex organic compounds, such as carbohydrates. This unit starts by looking at different types of **heterotrophic nutrition**, including saprobiontic and parasitic nutrition.

Autotrophic organisms are able to synthesise their own organic substances from simple inorganic compounds, using light energy, in the process of **photosynthesis**. In this way, light energy is trapped and made available to heterotrophic organisms in the form of chemical energy of the organic substances. In ecological studies, autotrophic organisms are often referred to as *producers*; energy flows from producers to consumers, through **food chains** and **food webs**.

Unit 3 also includes some examples of the ways in which elements such as **carbon** and **nitrogen** are cycled in ecosystems and how human activities may disrupt nutrient cycles. You will also study energy resources, including the use of **fossil fuels**, **biogas** and **gasohol**, and the impact of human activities on the environment with reference to **deforestation**, **desertification**, **atmospheric pollution** and **water pollution**.

A good knowledge and understanding of this unit will provide you with a firm foundation for studying other aspects of ecology.

The unit is divided into two parts, with six topics in Part a:

Part a
1 **Modes of nutrition**
2 **Ecosystems**
3 **Energy flow**
4 **Recycling of nutrients**
5 **Energy resources**
6 **Human influences on the environment**

Part b
The individual investigation

Unit 3

Part a: Topics

Topic ① Modes of nutrition

Introduction

It is desirable to have a basic understanding of the different modes of **nutrition** in living organisms in order to appreciate the relationships between plants and animals and to understand the **flow of energy through ecosystems**. You will already have some knowledge of digestion in humans from the section in Unit 2 and an outline of the roles of producers and consumers from your studies at GCSE or the equivalent. The details of digestion are not required in this unit. Knowledge of **autotrophic**, **holozoic**, **saprobiontic**, **parasitic** and **mutualistic** nutrition is required. It is necessary to know the characteristic features of each of these together with specific adaptations shown by named organisms.

A knowledge of the roles of biologically important molecules (Unit 1) helps in understanding the different types of nutrition and some understanding of exchange processes and the transport of nutrients around the body (Unit 2) is also helpful. It should be stressed that a clear understanding of all the different terms used is necessary and it would be wise to learn a definition for each.

It should be noted that you do not need to know the details of photosynthesis at this stage, but you should appreciate the major differences between autotrophic and heterotrophic nutrition.

Checklist of things to know and understand

Before attempting to answer any of the questions, check that you know and understand the following:

- [ ] the basic principles of autotrophic and heterotrophic nutrition

- [ ] that holozoic nutrition involves feeding on organic matter from the bodies of other organisms

- [ ] the adaptations of herbivores to their mode of nutrition as illustrated by a named ruminant

- [ ] the adaptations of carnivores to their mode of nutrition as illustrated by a named example

- [ ] saprobiontic nutrition as illustrated by *Rhizopus*

- [ ] parasitic nutrition as illustrated by *Taenia*

- [ ] mutualistic nutrition as illustrated by *Rhizobium* with Papilionaceae and cellulose-digesting organisms in ruminants.

Testing your knowledge and understanding

To test your knowledge and understanding of modes of nutrition, try answering the following questions.

1.1   For each description given below, identify the mode of nutrition

---

 The answers to the numbered questions are given on pages 86–103.

and then name an example:
- **(a)** Organisms which obtain their food from the bodies of other organisms
- **(b)** Organisms which synthesise organic molecules from inorganic raw materials using light energy
- **(c)** Organisms which feed on dead or decaying organic matter
- **(d)** Organisms which feed only on plant material
- **(e)** Organisms which obtain food from another living organism causing deprivation or damage to the host organism
- **(f)** Organisms which live in partnership with another living organism, both benefiting from the association
- **(g)** Organisms which are secondary consumers and feed on other animals

1.2 Name the major taxonomic groups to which each of the following organisms belongs:
- **(a)** *Rhizopus*
- **(b)** *Taenia*
- **(c)** *Rhizobium*

1.3 Which of the following features of teeth and jaws are adaptations of herbivores to their diet and which are adaptations shown by carnivores?
- **(a)** Long, pointed canine teeth
- **(b)** Open roots
- **(c)** Inwardly curved incisors
- **(d)** Diastema
- **(e)** Large surface area of molars
- **(f)** Enamel ridges on molars
- **(g)** Carnassial teeth
- **(h)** Horny pad on upper jaw

You could extend this by giving a function of each of the features in the list above.

1.4 Test your knowledge of saprobiontic and parasitic nutrition. Make diagrams of *Rhizopus* and *Taenia* from memory, check to see that you are right and then label all the features. You could annotate the features to show how they are related to the mode of nutrition of the organism. This type of exercise would be useful on a revision card.

1.5 The specification states that you should know the details of the relationship of mutualistic nutrition between *Rhizobium* and members of the plant family Papilionaceae (pea family). It is worth noting that this relationship is important in the nitrogen cycle and the way in which nitrogen is made available for plant growth, so will be relevant in later sections of this unit. You also need to know about the organisms which play a role in cellulose breakdown in the ruminant gut.

Try drawing up a balance sheet, so that you can see what each organism gains and what it contributes to the association in which it is involved.

see Activities toolkit, page xii.

Helpful hints

Where you are left to choose which ruminant and which carnivore to study, it would be wise to choose examples which are dealt with in detail in textbooks. The dentition of sheep, cats and dogs is well documented in most textbooks, but there is less information about the rest of the alimentary canal. Diagrams of the digestive systems of ruminants, such as the cow, are not difficult to find and carnivores' guts are very similar to our own.

Mark allocations are given for each part of the question and the answers are given on pages 86–103.

Practice question

1   The table below refers to parts of the alimentary canal of a cow and their functions.

Complete the table by writing in an appropriate function for each part.

Part of alimentary canal	Function
abomasum	
omasum	
reticulum	
rumen	

(Total 4 marks)

(Modified from Edexcel B / HB4C, January 1997, Q. 2)

Topic **2,3** Ecosystems and energy flow

Introduction

These two topics cover the components that make up ecosystems and how they interrelate. You need to know that ecosystems harness energy by **photosynthesis** and that energy is passed on as well as lost from **food chains**. You will be able to apply what you have learnt about **exchanges with the environment**, in Unit 2 and about different **modes of nutrition** from this unit. You need to remember that carbon dioxide and water are converted to glucose and oxygen during photosynthesis. The parts of an ecosystem can be divided into abiotic and biotic components. The links in a **food web** are where an organism occurs in more than one food chain. A **pyramid** is usually a series of rectangular blocks. The area of each block is equivalent to the number, **biomass** or energy of organisms at one trophic level.

Always look at the units given on a diagram of a pyramid. The units on a pyramid of biomass for a terrestrial ecosystem will usually be kg m^{-2} because it is related to the area of land. For an aquatic ecosystem the units will usually be kg m^{-3} because the pyramid relates to the volume of the water.

Knowledge of this section will be useful when you study how humans influence the structure of ecosystems in this unit.

Checklist of things to know and understand

Before attempting to answer any of the questions, check that you know and understand the following:

☐ the terms *biosphere*, *ecosystem* and *habitat*

- [ ] that sunlight energy is absorbed by chlorophyll and used in photosynthesis
- [ ] the terms *gross primary productivity* and *net primary productivity*
- [ ] the role of producers in gross primary productivity and net primary productivity
- [ ] that trophic levels are made up of producers or consumers or decomposers
- [ ] the role of producers, consumers and decomposers in food chains and food webs
- [ ] that a food web is composed of food chains
- [ ] how energy passes through food chains and food webs
- [ ] why energy is lost at each trophic level
- [ ] how to construct pyramids of numbers, biomass and energy for a food chain.

Practicals

You are expected to have carried out practical work to investigate:

- [ ] *the fresh biomass of trophic levels*
- [ ] *the numbers and fresh biomass of organisms at each trophic level and to use this to construct pyramids of numbers and fresh biomass.*

 *Practical work – Helpful hints*

In this section, you are expected to understand techniques for the collection of biomass and to construct pyramids of fresh biomass and numbers.

- Biomass is defined as the mass of living material per unit area or unit volume.
- When collecting biomass, you should use a standardised collection method, such as a quadrat for terrestrial habitats, or a bottomless dustbin and sweep net for aquatic habitats.
- Sort out the organisms into trophic levels – producers, primary consumers (herbivores) and secondary consumers (carnivores) – and weigh, to find fresh biomass, or count the total number of organisms in each trophic level. Record your results in a suitable table.
- Return the organisms, unharmed, to their habitat.
- When drawing ecological pyramids remember the following rules.
- Arrange the bars with producers at the bottom, then at successive trophic levels.
- The width of the bars is proportional to the numbers, biomass, or energy.
- The pyramid should be symmetrical.
- Include a scale and label the trophic levels.

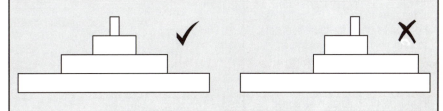

Unit 3

Testing your knowledge and understanding

The answers to the numbered questions are on pages 86–103.

Helpful hints

You could expand this into a complete vocabulary list, with brief explanations and use them as revision cards. See Activities toolkit, page xii.

To test your knowledge and understanding of ecosystems and energy flow, try answering the following questions.

2, 3.1 List *five* abiotic factors that could affect organisms in an ecosystem.

2, 3.2 Write down the meaning of the following terms: *biosphere, habitat* and *consumer*.

2, 3.3 Write down the equation for photosynthesis.

2, 3.4 Sketch a pyramid of number to represent this food chain.

oak tree ⇒ caterpillars ⇒ starlings ⇒ hawk

2, 3.5 List *five* ways in which energy is lost at a trophic level.

2, 3.6 Draw up a table that has the following headings: 'Producers', 'Primary consumers' and 'Secondary consumers' and complete it using the names of the organisms shown in the food web below.

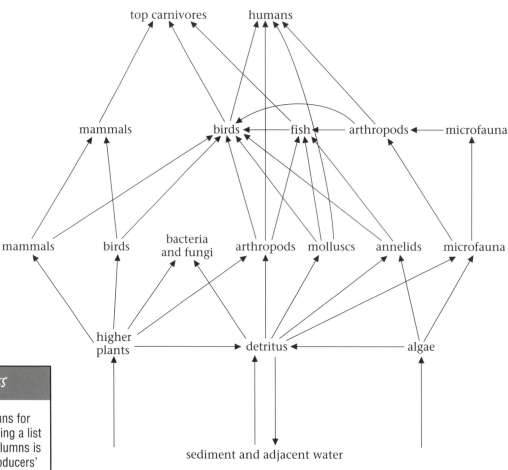

Helpful hints

You could add further columns for the other trophic levels. Adding a list of features to each of the columns is a good idea, e.g. for the 'Producers' column, use sunlight, inorganic molecules, synthesis of organic molecules, autotrophic. How biomass or energy is lost from each trophic level could also be added.

2, 3.7 List the stages needed in a practical investigation that will enable a pyramid of fresh biomass for a pond to be constructed.

Practice questions

1 Explain the differences between the following pairs of terms.
   (a) *Food chain* and *food web*   [2]
   (b) *Producer* and *decomposer*   [2]
   (c) *Community* and *ecosystem*   [2]
                                                               (Total 6 marks)
                                                               (*New question*)

2 The diagram shows the energy flow for a large pond. All values are given in kJ m^{-2} yr^{-1}.

Sunlight
⇓ 3×10^6

| Phytoplankton (microscopic plants) GPP = 2.8×10^4 NPP = 1.8×10^4 | 4×10^3 ⇒ | Zooplankton (microscopic animals) | 400 ⇒ | Small fish |

Note: GPP = gross primary production. NPP = net primary production.

   (a) Name the trophic levels to which each of the following belong.
       (i)   Zooplankton   [1]
       (ii) Small fish   [1]
   (b) (i)   Calculate the percentage energy from sunlight which is
             fixed as GPP by phytoplankton. Show your working.   [2]
       (ii) Suggest *two* reasons why not all of the incident sunlight
             is utilised in photosynthesis.   [2]
                                                               (Total 6 marks)
                                                               (*ULEAC, June 1993, Q. 4*)

Topic ④   Recycling of nutrients

Introduction

In order to understand how water, carbon and nitrogen are recycled through ecosystems you need to revisit some of the content of Unit 1. Your notes from this unit will have information about the properties of water and its roles in living organisms. Information about the types of compounds that contain carbon and nitrogen will also be there. You will also need to remember some of the processes that occur in living organisms, such as **respiration**, **photosynthesis**, **gas exchange** and **transpiration**. Try to remember the parts of the cycles by thinking of the processes, the compounds involved and the organisms. Start with nitrogen gas in the air for the **nitrogen cycle**, and carbon dioxide gas in the air for the **carbon cycle**. Build up your own diagram from these starting points. Trying to memorise one from a textbook is always difficult. Nitrifying, nitrogen-fixing and denitrifying are often confused, so try to get the differences sorted out in your mind before you start to learn the whole of the nitrogen cycle.

What you have learnt in this section will also be helpful when you look at the impact of human activity on the environment in this unit.

Checklist of things to know and understand

Before attempting to answer any of the questions, check that you know and understand the following:

- ☐ the stages in the water cycle

- ☐ the stages in the carbon cycle

- ☐ the role of microorganisms, carbon sinks and carbonates in the carbon cycle

- ☐ the stages in the nitrogen cycle

- ☐ the role of nitrifying bacteria, denitrifying bacteria, nitrogen-fixing bacteria and decomposers in the nitrogen cycle

- ☐ how human activities disrupt the carbon and nitrogen cycles.

Testing your knowledge and understanding

To test your knowledge and understanding of recycling of nutrients, try answering the following questions.

4.1   List *four* of the physical processes involved in the water cycle.

4.2   Describe how the following are involved in the cycling of carbon: respiration, photosynthesis, carbon sinks.

4.3   Starting with nitrogen gas draw the part of the nitrogen cycle that involves microorganisms.

The answers to the numbered questions are on pages 86–103.

Helpful hints

Once you have found a way of learning each of the cycles it would be a good idea to put each of them onto a separate revision card. You could highlight each part in a different colour to help you to recall what is on the card. Then when you are searching your memory for an answer in the exam you should be able to picture the card in your head. See Activities toolkit, page xii.

4.4 Using the diagram of the carbon cycle shown below, draw out and complete the table with all the letters shown.

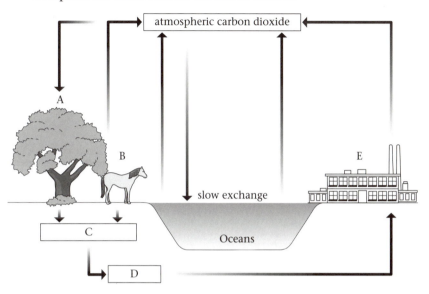

Letter	Name of process	Conditions needed
A		
B		
C		
D		
E		

Helpful hints

You could add another column to this table that included the names of organisms involved in each process. If the same exercise is done for each of the cycles then you could use them to test yourself when revising this section.

see Activities toolkit, page xi.

4.5 Draw a spider diagram to show how the human activities of burning fossil fuels, cutting down trees and production of methane, can have an impact on the carbon cycle. An outline for you to use is given below.

Helpful hints

The impact of the human activities of fertiliser use, use of detergents and disposal of sewage and their impact on the nitrogen cycle could be shown in a similar way.

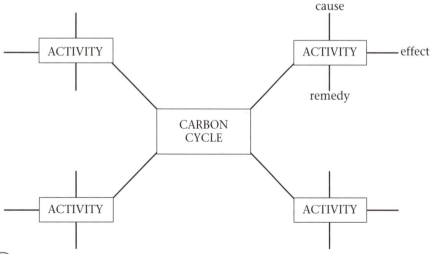

Mark allocations are given for each part of the question and the answers are given on pages 86–103.

Practice question

1 The sequence shown overleaf shows stages in the cycling of nitrogen in natural systems.

Unit 3

```
                    Nitrogen in                Denitrification
                    atmosphere
         ┌──────────────────────────────────────────────────────────┐
         │                                                            │
         │        Protein in         Eaten    Protein in             │
         │    ┌─▶ plants and  ─────────────▶  animals                │
         │    │   microbes                                            │
    A    │    │        │ Death                     │ Death,          │
         │ Assimilation│                           │ nitrogenous     │
         │    │   Protein in                       │ excretion,      │
         │    │   dead organic  ◀──────────────────┘ e.g. urea,      │
         │    │   matter + excreta                   defaecation     │
         │    │        │              C                              │
         │    │        ▼                                             │
         │    └── NH₃, ammonia/                   Denitrifying       │
         └─────── NH₄⁺, ammonium                  bacteria, e.g.     │
                  compounds                        Thiobacillus      │
              ┌──────┐                             denitrificans,    │
     Absorption│      │                      ⎫    Pseudomonas        │
     e.g. by roots  NO₂⁻, nitrites           ⎬ B  denitrificans      │
              │                               ⎭                      │
              │        │                                             │
              │        ▼                                             │
              └── NO₃⁻, nitrates ──────────────────────────────────┘
```

(a) Name the processes represented by arrows A, B and C. **[3]**

(b) Describe the biological consequences that might arise as a result of large quantities of nitrate draining from agricultural land into a freshwater lake. **[4]**

(Total 7 marks)

(ULEAC, June 1994, Q. 5)

Topic ⑤ Energy resources

Introduction

In this topic you are looking at issues relating to energy use – the ways people use energy and their sources of fuel. As we move into the twenty-first century, you should try to see this in relation to the past, the present and the future. You need to think about the conflict between continued (and increasing) use of fossil fuels and the growing trend to utilise **renewable resources**. Think also about the reasons why we use different energy sources, their origins and some of the consequences of using these fuels. You need to consider 'traditional' energy sources together with current sources, such as electricity and motor fuels, and to look at these in relation to the situation in developing nations as well as in industrialised nations. You will look at some specific examples of 'biofuels' and consider their potential contribution to fuels in the future.

In some respects you are dealing with big issues for which there are no simple answers just to 'learn'. In trying to find answers to questions that are raised, you may need to consider the availability of fuels, and also economic, social or perhaps political factors. Overall, you should relate this topic to environmental issues, on a local as well as a global scale and it is important that you apply your *biological* knowledge to help you understand these wider issues.

You should link your revision in this topic to other topics in Unit 3. You need to understand about the role of **photosynthesis** in the production of **biomass**, and also to be familiar with events in the **carbon cycle**. This builds on work you will have done in your GCSE or equivalent before starting AS. You should also refer to the processes of respiration (Unit 2) and be aware that some **fermentations** are carried out by yeasts and bacteria. You will certainly link this topic to the **environmental impact of human activities**, particularly those related to **air pollution**, including **acid rain** and the **enhanced greenhouse effect**.

 Checklist of things to know and understand

Before attempting to answer any of the questions, check that you know and understand the following:

- ☐ how the fossil fuels coal and oil were formed

- ☐ some environmental effects of burning fossil fuels

- ☐ the reasons for encouraging the practice of utilising renewable energy sources

- ☐ how fast-growing biomass can be utilised as a renewable energy source

- ☐ how gasohol and biogas are produced

- ☐ how energy resources can be managed in a sustainable manner.

 Testing your knowledge and understanding

To test your knowledge and understanding of energy resources, try answering the following questions.

5.1 The table below summarises trends in global energy use, by source, from 1990 to 1998. Use this as a stimulus, and the questions below it, to look at the key issues relating to energy use.

Energy source	Annual rate of growth (%)
wind power	22.2
solar photovoltaics	15.9
geothermal power	4.3
hydroelectric power	1.9
oil	1.8
natural gas	1.6
nuclear power	0.6
coal	0.0
Source: *Vital Signs – 99/00*. Earthscan Publication.	

(a) Which of the energy sources (listed in the table) are fossil fuels?
(b) Describe how *two* of these fossil fuels are formed.
(c) Which of these energy sources (listed in the table) can be included under 'renewable' energy sources?
(d) Suggest why wind power and solar photovoltaics (cells which generate electricity directly from solar radiation) showed the highest annual growth over this period.

 The answers to the numbered questions are on pages 86–103.

Helpful hints

Before you answer this, make sure you are clear about your definition for 'renewable' energy. (You would need to do this in a similar way when answering a question in a unit test.) A simple definition is *energy sources which are replenished at the same rate as they are used*. Another definition, as recently defined by the UK Renewable Energy Advisory Group, is *those energy flows that occur naturally and repeatedly in the environment and can be harnessed for human benefit. The ultimate sources of most of this energy are the sun, gravity and the Earth's rotation*.

Unit 3

Unit 3

Helpful hints

To answer this you need to know about the NFFO (Non-Fossil Fuel Obligation). This was part of the 1989 Energy Act in the UK. Its aim was to encourage development of ways of generating electricity which did not increase atmospheric carbon dioxide and also to promote diversification of energy supplies, particularly those derived from renewable energy sources.

5.2 Living biomass can be converted into energy to provide a renewable energy source. In recent years, the term *energy crops* has been used to describe plants grown specifically as energy sources.
   (a) Name *two* energy crops grown in the UK.
   (b) Name *one* or *more* energy crops grown outside the UK.
   (c) Choose *one* example (such as short-rotation coppicing) and describe how the crop is grown and harvested.
   (d) Explain how these energy crops can contribute to meeting requirements of UK legislation, such as the NFFO (Non-Fossil Fuel Obligation).

5.3 **Gasohol** is a fuel produced from biomass.
   (a) What is the composition of gasohol and what is it used for?
   (b) What source material(s) is (are) used to produce the ethanol in gasohol?
   (c) Name the biological process which produces ethanol.

5.4 **Biogas** is a fuel produced from biomass.
   (a) What is the approximate composition of biogas and what is it used for?
   (b) What source materials are used to produce biogas?
   (c) Name the organisms involved in the process which produces methane and list the main stages of the process.

5.5 These (energy crops, gasohol and biogas) and other fuels from biomass are described as 'renewable' fuels.
   (a) Apart from their use as an energy source, what other values do these fuels have? List any useful by-products obtained from these fuels and think about possible environmental benefits. Are there any *disadvantages* to using renewable fuels?
   (b) In questions 5.2, 5.3 and 5.4 you looked at energy crops, and materials which are used to produce gasohol and biogas. List some other sources of biomass that are used for fuels. Indicate whether they are burnt directly or converted into a combustible fuel.
   (c) For each of these fuels, try to decide whether its production can be managed sustainably.

5.6 **Landfill gas** can be used as an energy source.
   (a) What is landfill gas? What happens to it if it is *not* harnessed as a fuel source?
   (b) After closure of a landfill site, over what time-scale can the landfill gas be harnessed?
   (c) How does landfill gas fit into our consideration of renewable fuels? If we use it as a fuel, how does this help with environmental issues, such as the enhanced greenhouse effect?

When you have completed the answers to these six questions, you should find that you have made a very useful summary of the details of different parts of this topic on energy resources. Now go back to the introduction and think again about the 'big issues' related to energy resources and try to identify them. In particular, see if you have improved in your answers to the following two points from the initial checklist: (1) the reasons for encouraging the practice of utilising renewable energy sources, and (2) how energy resources can be managed in a sustainable manner.

Practice question

1 The diagram below shows the outline of a process for the production of gasohol.

Mark allocations are given for each part of the question and the answers are given on pages 86–103.

Corn extract   Amylase   Yeast

| Fermenter | ⇒ | Distillation | ⇒ | Gasohol |

(a) Explain why amylase is added to the fermenter. **[2]**
(b) Name *one* constituent of gasohol. **[1]**
(c) Give *two* benefits of using gasohol, rather than petrol, as fuel. **[2]**
(d) Suggest *one* product of economic importance, other than gasohol, which could be obtained from the fermenter. **[1]**

(Total 6 marks)
(*Edexcel HB2, June 1998, Q. 4*)

Topic ❻ Human influences on the environment

 Introduction

In this topic you are looking at issues relating to the environment. It is really a 'cause' and 'effect' section. The *causes* can all be traced back to human activities and the *effects* result in harm, damage or changes in the environment, in ecosystems and the living organisms within them. In a similar way to Topic 5 (*Energy resources*) you are dealing with big issues for which there are no simple answers just to 'learn'. In trying to understand the implications of the events which occur (or have occurred), and to find answers to questions that are raised, you should apply your biological knowledge but also consider economic, social or perhaps political factors. You should be able to relate this topic to events on a local as well as a global scale and always try to find specific examples which help to illustrate the principles you are exploring. You do need to make sure you present your answers in a firmly scientific framework and avoid slipping into broad sweeping generalisations that are biologically inaccurate. As a biologist studying for AS, you should not be trying to imitate a sensationalist media correspondent.

It is sensible for you to do your revision for this topic in two parts. The first would look at **deforestation** and **desertification** – and here it may help you to compare one with the other and find parallel points to consider. The second part looks at the topics of **air pollution** and **water pollution**. Remember that you are looking for causes and effects – and always, for ways to reverse the adverse trends and recovery from damage caused. You should link your revision here to other topics in your AS course. For deforestation and desertification, you need to have an understanding of **ecosystems**, and have some appreciation of **biodiversity** and how much biomass can be supported on a particular area of land. You would also find it helpful to know about **adaptations** shown by plants in extreme dry conditions (see Unit 2). For your revision on air pollution and water pollution, you need to know

Unit 3

about the **burning of fossil fuels** (see Topic 5) and have an understanding of the **ecology of aquatic organisms** (see Unit 2) as well as some of the adaptations shown by them to different conditions.

Deforestation and desertification

 Checklist of things to know and understand

Before attempting to answer any of the questions, check that you know and understand the following:

☐   why deforestation occurs and some idea of its extent

☐   how the deforested area becomes degraded

☐   the wider effects of deforestation – on the atmosphere, nearby communities and biodiversity

☐   how measures can be taken to allow recovery of a deforested area

☐   why desertification occurs and how it is linked to human activities

☐   how degradation of soil in the area contributes to the process of desertification

☐   how measures can be taken to allow recovery of an area which has undergone desertification.

 Testing your knowledge and understanding

 The answers to the numbered questions are on pages 86–103.

To test your knowledge and understanding of deforestation and desertification, try answering the following questions.

Deforestation

During the last decade of the twentieth century, the annual loss in forest cover was about 16 million hectares (per year) and globally over 100 countries suffered a net loss in forest. As well as loss in terms of quantity, there were changes in the nature of the forests in reforested areas.

6.1   List some reasons why people cut down forests.

6.2   Is deforestation just a twentieth-century event, or has it been going on for a long time? Give times, dates or some examples to back up your answer.

6.3   When deforestation occurs, describe what is likely to happen to each of the following: the soil; rivers in the area; the nutrient balance in the soil; the oxygen / carbon dioxide balance in the atmosphere; the biodiversity.

6.4   Explain how erosion can occur in a deforested area.

6.5   Give some examples of changes in species diversity in a deforested area.

6.6   How far do you think that the practice of coppicing can be considered a means of sustainable management of a woodland (or forest)? List key points that would support your answer.

6.7   Suggest other ways that forests could be managed sustainably.

6.8   Check your definitions: *deforestation; afforestation; reforestation.*

Desertification

Desertification can have serious consequences for people living in an already fragile area, with threats of famine and of starvation.

6.9 List some reasons why desertification occurs. Make it clear how the process is related to activities of people. You should also try to link desertification to the *carrying capacity* of the land.

6.10 In an area affected by desertification, explain the events that lead to each of the following: erosion; salinisation.

6.11 Explain how poor irrigation can increase salinisation.

6.12 Suggest ways that an area could be *managed* to allow it to recover from desertification. Then explain how the area can be maintained to allow production of crops or be used as grazing land. Give some examples to support your answer.

Atmospheric and water pollution

 Checklist of things to know and understand

Before attempting to answer any of the questions, check that you know and understand the following:

❑ which gases contribute to acid rain and how acid rain is formed

❑ some examples of the effects of acid rain on living organisms

❑ how measures can be taken (by individual people and their governments) to reduce the output of gases which lead to the formation of acid rain

❑ how the existing greenhouse effect influences conditions on Earth

❑ which gases contribute to the greenhouse effect

❑ how these greenhouse gases may be produced by human activities

❑ some of the possible consequences of the *enhanced* greenhouse effect, resulting from human activities

❑ how measures can be taken (by individual people and their governments) to reduce the output of greenhouse gases

❑ how changes in oxygen, carbon dioxide, light intensity and nutrients affect aquatic organisms

❑ that raw sewage and excess fertiliser can affect water quality

❑ how eutrophication can lead to algal blooms

❑ the effects of algal blooms on other aquatic organisms

❑ how measures can be taken (by individual people and their governments) to maintain good water quality or improve water quality following the effects of pollution

❑ how legislation can monitor, control and reduce the effects of pollution to maintain desired quality of the atmosphere and of water.

The answers to the numbered questions are on pages 86–103.

Testing your knowledge and understanding

To test your knowledge and understanding of atmospheric and water pollution, try answering the following questions.

Atmospheric pollution

6.13 Look at the graphs below which show carbon emissions from fossil fuel burning, during the years 1950 to 1998. Assume these graphs represent all the countries in the world.

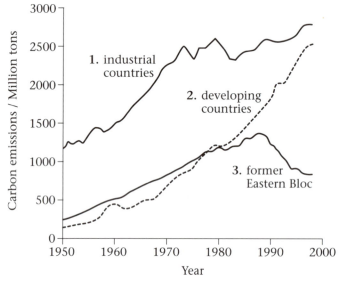

Source: *Vital Signs – 99/00*. Earthscan Publication.

Helpful hints

In doing this, make sure you relate the rises and falls to actual dates. This advice applies to all these data questions or wherever you are looking at a trend. Quote actual figures to support your answers.

(a) Describe the shape of each of these three graphs (1. industrial countries, 2. developing countries and 3. former Eastern Bloc).

(b) Then try to give reasons for the trends and changes – in particular, look at the patterns from 1980 to the late 1990s.

(c) From the three graphs, work out the total *world* carbon emissions from fossil fuel burning over the same period. You can read off the figures at 10-yearly intervals then plot your totals on a graph.

6.14 The graph on the following page gives changes in atmospheric concentration of carbon dioxide, at 40-year intervals, from 1760 to the present day.

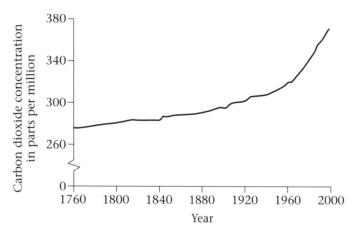

Source: *Vital Signs – 99/00*. Earthscan Publication.

(a) Take 1760 as a 'pre-industrial' level, and work out the percentage increase over successive 40-year intervals. First you can work out the *actual* increase (in parts per million), compared with the pre-industrial level.

(b) What links can you make between the carbon emission data in question 6.13 and the atmospheric carbon dioxide concentration?

(c) Explain how increases in atmospheric carbon dioxide affect the greenhouse effect.

(d) What gases, other than carbon dioxide, are given off from the combustion of fossil fuels?

6.15 Sulphur dioxide is one of the gases which contributes to the formation of **acid rain**. The table below gives data about sulphur dioxide emissions from fossil fuel burning. It refers to three areas of the world, Europe, the USA and Asia, and covers the years 1980 to 2010. Data for the years 2000 and 2010 are predictions, based on models.

| Year | Sulphur dioxide / million tons | | |
	Europe	United States	Asia
1980	59	24	15
1990	42	20	34
1995	31	16	40
2000	26	15	53
2010	18	14	79

Source: *Vital Signs – 98/99*. Earthscan Publication.
[Note: 1 ton is approximately equal to 1 tonne (metric ton)]

(a) Plot or sketch a graph of these data and describe the trend shown by each of these areas.

(b) Suggest reasons for the differences in the trend shown by Europe and that shown by Asia. Remember to link your reasons to the actual data and to dates shown in the table.

Unit 3

see Activities toolkit, page xii.

Helpful hints

Avoid vague answers – like 'acid rain kills plants . . . ' – you will not get many marks in a unit test answer.

There are concerns that human (anthropogenic) activities are now responsible for an *enhanced* greenhouse effect. This may have both long term and far reaching effects on climate, particularly in relation to global warming. Some of these changes may be irreversible and escalating.

In a glasshouse (greenhouse), the glass is transparent to incoming radiation, inside the air and ground warm up and the panes of glass help to retain heat – partly because they prevent warm air escaping.

There are at least five gases that we describe as 'greenhouse gases' because they are considered to contribute to the greenhouse effect. Some of these arise from anthropogenic activities, while some originate from other sources.

6.16 Draw a simple flow chart to summarise the events which take place after the release of sulphur dioxide from burning of fossil fuels until its deposition as acid rain.

6.17 Name *two* gases, other than sulphur dioxide, which contribute to the formation of acid rain.

6.18 Summarise some of the effects of acid rain on living organisms. Make sure you give specific examples, and try to identify the effect in detail. You could write your answers in a table. The organisms could include phytoplankton, fish and other aquatic species, conifers and deciduous trees and you could also list some effects on human health.

6.19 The **greenhouse effect** is a natural phenomenon which has a profound influence on the Earth as we know it and live in it. What sort of temperatures might there be on Earth without the existing greenhouse effect? How would these temperatures affect people, their lives and the crops that could be grown?

6.20 Using the analogy of a glasshouse (greenhouse), try to explain, in simple terms, how the atmosphere around the Earth creates the 'greenhouse effect'. Use these short questions to help you. What is the atmosphere equivalent to? How is UV radiation 'filtered out' of the incoming radiation? How is heat energy re-radiated back to Earth so that it is 'retained'? Where do 'greenhouse gases' accumulate to enhance this effect?

6.21 The table below lists either greenhouse gases or their likely origin. Complete the table so that you have a summary of information about the most important greenhouse gases. Then arrange the gases in the order of their effectiveness, relative to carbon dioxide.

Name of greenhouse gas	Origin or source
carbon dioxide	
	aerosols; refrigerators;
	guts of livestock (such as ruminants); anaerobic bacteria; rotting vegetation;
nitrous (and other) oxides (NO_x)	
ozone	

6.22 What evidence is there that the global warming being experienced now, and as predicted into the twenty-first century, is a consequence of human (anthropogenic) activities?

6.23 List changes that are likely to take place as a result of the enhanced greenhouse effect, and the resulting global warming. Indicate possible benefits (to people or other living organisms) as well as the harmful effects.

6.24 List some measures that can be taken to limit emission of greenhouse gases. Include some things that you or other people can do at an individual level as well as attempts or agreements made at a local, national or international level.

6.25 Check your definitions: *greenhouse effect, global warming.*

Water pollution

6.26 The table below lists *three* key factors which can change in a freshwater aquatic environment. Work through the table and find answers to the points listed. You then have a useful summary which should help you understand the main issues related to water pollution.

To understand the effects of pollutants in water, you need to think first about some of the factors in an aquatic ecosystem – how they interact with the aquatic organisms and how they influence which organisms are found there. Then you can apply this information to a situation where water pollution has occurred – say as a result of raw sewage being discharged into a river.

Factor	Questions to answer
oxygen	**(a)** How does the solubility of oxygen in water change with increases in temperature? Describe some adaptations shown by aquatic organisms in very *low* levels of oxygen. Give reasons why water may become depleted of oxygen. List ways of increasing oxygenation in water – both artificially and naturally.
carbon dioxide	**(b)** How does the solubility of carbon dioxide in water change with increases in temperature? How may the pH of water change – during 24 hours, and at different seasons in temperate regions? Explain likely fluctuations in relation to carbon dioxide levels.
light intensity	**(c)** How does light intensity change with depth in water? List possible suspended particles, both non-living and living, that might be present and which can alter light intensity in water. What effect might such suspended particles have on photosynthetic organisms in water?

Remember that, when describing nutrients in water, reference is made mainly to inorganic nitrogen and phosphorus, though other nutrients are, of course, important to the organisms in the water.

The **nutrient** level in a body of water depends partly on the bedrock and the soil in the area. It may also be affected by anthropogenic activities.

Work through the following questions to help you to understand what happens when there are sudden changes in the nutrient status in water, due usually to an excess of nutrient, otherwise known as *eutrophication*. You can then link these events to the origins and effects of pollution.

6.27 Check your definitions: *oligotrophic* and *eutrophic*; *phytoplankton* and *zooplankton*.

6.28 Name at least *three* ways in which human activities can lead to increases in the nutrient level in water.

Helpful hints

Look at the parts of the words to help work out and remember their meaning – *oligo* = few, *eu* = plenty, *phyto* = plant, *zoo* = animal – and you should be able to work out 'trophic' and 'plankton' from your other work. What other words can you think of with these same 'parts'?

6.29 (a) Following a sudden increase in nutrient, explain why there may be a sudden increase (population explosion) of algae, otherwise known as an *algal bloom*.
(b) Then explain why the algal bloom may suddenly collapse and lead to a *depletion* of oxygen.
(c) How would the increased algae (living and dead) affect light intensity in the water and what effect would this have on other photosynthetic organisms?

6.30 What does the term *fish kill* mean? At what time of day is it most likely to occur? Explain your answer.

6.31 List some other harmful effects arising from eutrophication.

6.32 Consider a situation when raw sewage is accidentally discharged into a river, which was relatively clean. The water becomes polluted. In the zone downstream from the polluted zone, changes take place in the water and in the organisms that are found there. Still further downstream, the water becomes clean again.

(a) Complete the chart below to help you trace the events which occur in the water. In most cases, you can choose from the following words – high / falls / rises / low – to represent the changes. The chart starts in a zone of clean water, just above the sewage discharge. The descriptions in the first column summarise stages in recovery of the water. The double line indicates where the sewage entered the water.

Zones downstream (*above and below discharge*)	Oxygen	BOD	Dissolved solids Suspended solids	Ammonia	Nitrates	Phosphates
clean water						
degradation						
active decomposition						
recovery						
clean water						

(b) Where, in these zones, would you find *peaks* of the following microorganisms? algae; bacteria; *Cladophora*; protozoa / protoctista; sewage fungi;

(c) Where, in these zones, would you find *peaks* of the following animals and roughly where would the increases start? For each of the following animals (or animal group), identify features which enable them to live in the conditions found there. *Asellus*; *Chironomus*; clean water fauna; Tubificidae;

Practice questions

1 Acid rain is a matter of serious environmental concern. Sulphuric acid is present in acid rain and has adverse effects on both plants and animals.

(a) (i)   Name *two* acidic components of acid rain other than sulphuric acid.   **[2]**

     (ii)   Describe how acid rain is formed.   **[3]**

(b) An experiment was carried out to investigate the effect of dilute sulphuric acid on the growth of cress seedlings. Batches of seeds were sown in glass dishes on filter paper to which dilute sulphuric acid was added. The dishes were then incubated. The root and shoot lengths were measured after 65 hours. The results are shown in the table on the following page.

Mark allocations are given for each part of the questions and the answers are given on pages 86–103.

Sulphuric acid concentration / mol dm^{-3}	Mean root length / mm	Mean shoot length / mm
0	55.5	25.2
1×10^{-3}	63.4	18.4
3×10^{-3}	6.5	9.5
4×10^{-3}	2.0	4.6
6×10^{-3}	2.8	0.8
7×10^{-3}	1.5	0.5
8×10^{-3}	1.3	0.3
9×10^{-3}	1.3	0.0
10×10^{-3}	1.0	0.0

(i) Describe the relationship between the concentration of sulphuric acid and the growth of roots as shown by the results in the table. **[2]**

(ii) Compare the effects of sulphuric acid on the growth of roots and shoots. **[3]**

(iii) Suggest *two* reasons why cress seedlings are suitable for investigating the effect of acid rain on plants. **[2]**

(Total 12 marks)
(Edexcel B2, January 1998, Q. 6)

2 Read through the following passage about pollution of freshwater by raw sewage, then write on the dotted lines the most appropriate word or words to complete the account.

> Sewage contains mineral ions such as nitrates and .... , and also suspended solids. If raw sewage flows into a river, the suspended solids are broken down by .... , resulting in a decrease in the concentration of dissolved oxygen in the water. The volume of oxygen used by a sample of water is known as the .... , which steadily .... as the organic solids start to be broken down. Mineral ions stimulate the growth of algae which can reduce the growth of submerged plants by reducing the amount of .... reaching them.

(Total 5 marks)
(Edexcel B2, January 1999, Q. 2)

 Mark allocations are given for each part of questions and the answers are given on pages 104–106.

Unit 3 Assessment questions

1 The diagram on the following page shows a longitudinal section through a molar tooth from a herbivorous mammal.

Unit 3

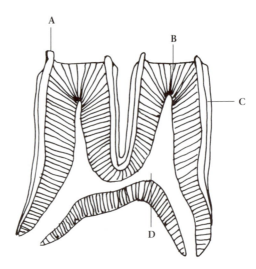

(a) Name the parts labelled A, B, C and D.   **[4]**

(b) Select *two* features shown on the diagram which are characteristic of herbivore dentition and for each feature, give *one* reason why it is an advantage to the herbivore.   **[4]**

(Total 8 marks)

(Modified from *Edexcel B / HB 4C, January 1996, Q. 1*)

2   The diagram below shows the quantity of energy flowing through a food chain in a terrestrial ecosystem. The figures are given in kJ m^{-2} yr^{-1}.

Incident sunlight
3×10^6
⇓

Green plants $\text{NPP} = 1.8 \times 10^4$	1800 ⇒	Caterpillars	100 ⇒	Insectivorous birds

(a) Calculate the percentage of the incident energy which becomes available as the net primary production (NPP) of green plants. Show your working.   **[2]**

(b) Give *two* reasons why not all the energy of the incident sunlight is incorporated into the biomass of green plants.   **[2]**

(c) Using the information shown in the diagram, explain why the biomass of insectivorous birds is usually very much less than the biomass of caterpillars.   **[2]**

(Total 6 marks)

(*Edexcel B2, June 1997, Q. 3*)

3 The diagram below shows an organism of the genus *Rhizopus*.

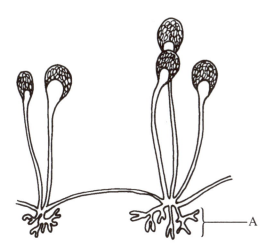

(a) Describe the role of the part labelled A in the nutrition of
the organism. **[3]**

(b) How does the nutrition of the tapeworm *Taenia* differ
from that of *Rhizopus*? **[3]**

(c) Explain what is meant by 'mutualistic nutrition'. **[2]**

(Total 8 marks)

(Modified from *Edexcel B2, June 1998, Q. 3*)

4 A group of students sampled the animal life in a pond. They plotted
their data as a pyramid of biomass as shown below. The biomass is
expressed as mg m^{-3}.

Biomass / mg m^{-3}

Tertiary consumers	3.5
Secondary consumers	26.8
Primary consumers	418.0

(a) Describe how the biomass values would have been
determined. **[3]**

(b) (i) Calculate the percentage decrease in biomass between
the primary consumers and the secondary consumers.
Show your working. **[2]**

(ii) Give *two* reasons for the decrease in biomass between
each trophic level. **[2]**

(Total 7 marks)

(*Edexcel B2, June 1999, Q. 3*)

5 Biogas is a mixture of gases produced by the action of anaerobic
bacteria on organic waste, such as animal manure. The bacterial
activity first produces carbon dioxide and organic acids. The organic
acids are then converted into methane. This process is carried out in
an enclosed container known as a digester.

(a) Give *two* reasons why an enclosed container is used for the
production of biogas. **[2]**

(b) The graph below shows changes in the pH in a biogas digester over a period of 60 days. The graph also shows the rate of gas production over the same period.

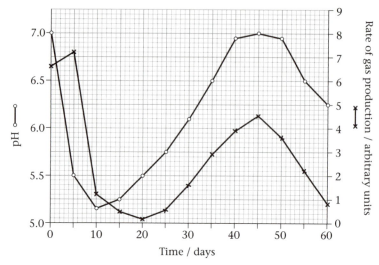

(i) Suggest explanations for the changes in pH in the digester during the following time periods:
  0 to 10 days
  10 to 45 days [3]

(ii) Suggest explanations for the changes in the rate of gas production during the period of 60 days. [3]

(c) Describe the advantages of using biogas, rather than fossil fuels, as a domestic fuel. [3]

(Total 11 marks)

(Edexcel B2, January 2000, Q. 6)

Part b:
The individual investigation

The individual investigation involves the assessment of practical skills through coursework. Once you have completed your investigation, it is marked by your teacher and then moderated (i.e. the marking is checked) by Edexcel. The aim of the investigation is to give you an opportunity to plan and carry out a scientific experiment on your own and it is meant to go a bit further than the practical work included in the specification. The investigation can be based on any of the AS units and it can be done in a laboratory or as part of your fieldwork. Your experiment should involve taking measurements so that you get quantitative (numerical) results and the practical work should take about three hours. You then need to write a detailed account of your investigation for your teacher to mark. It is very useful to have a copy of the marking scheme when you are planning, carrying out and writing up your investigation. If you know how the marks are awarded you are more likely to score highly.

The most important feature of the investigation is that it is individual. This means that it must be solely your own work and you cannot share results with a friend. Other students may investigate similar problems, but you are responsible for your own plan and, with the approval of your teacher, carrying out your own investigation. If you get in a muddle, your teacher can give you advice, but this is taken into account when the marks are awarded. It is sometimes better to seek advice at the beginning if you are really stuck at the planning stage, so that you can go on to gain reasonable marks for the other parts of the assessment.

In the individual investigation, marks are awarded for:

- planning
- implementing
- analysing evidence and drawing conclusions
- evaluating evidence and procedures.

For *planning*, you should explain what you are trying to investigate by formulating a hypothesis which you can test. It is often necessary to provide some biological background to the problem and explain the principles involved. You should design your experiment carefully, describing the apparatus you need to use and the measurements or observations you are going to make. This is crucial so that you can achieve useful and reliable results. At this stage, it is necessary to decide which factors or variables you need to control and how you will control them. You also need to consider the safety of your methods and the use of any living material or effects on the environment.

High marks can be achieved for *implementing*, provided that you can demonstrate that you handle all the materials and apparatus safely and competently, and that you are methodical in your approach to the investigation. The measurements and observations you make should be linked to your hypothesis. It is also very important that you take your measurements accurately and record them properly. All your original measurements should be recorded and it is often best to do this in the form of a table, using the correct units. For maximum reliability, it is a good idea to repeat your measurements or to use replicates in your investigation.

The section on *analysing evidence and drawing conclusions* gives you an opportunity to demonstrate your ability to present your results in the most suitable graphical form. Graphs should be chosen to display important trends, patterns or comparisons. Remember that you can often show trends and patterns more successfully on a summary graph than by using a number of separate graphs. If you are using a computer to generate your graphs, do look carefully at the axes and make sure that the points are shown clearly. In addition to the graphical representation of your results, you need to comment on the trends and draw some conclusions. Biological investigations do not always produce the results expected, but you should try and give a logical, coherent explanation of your findings, whatever they might be. These explanations should be related to the biological principles involved and any inconsistencies should be recognised and described.

Evaluating the evidence and the procedures gives you an opportunity to discuss the limitations of your methods and to note any difficulties that you encountered. You should be quite critical and show that you appreciate that you cannot always draw accurate conclusions from single investigations and that reliable results are sometimes very difficult to achieve. It is better to avoid excuses like 'if I had more time I would have . . . ', or 'the apparatus provided was not good enough . . . '. It is reasonable to suggest improvements that could be made to the technique and to suggest how the investigation could be taken further.

You should try to make your written account of the investigation as concise as possible without omitting important details. It is also important to make it legible and to pay attention to small details, such as the correct units on graphs and in tables. You should have plenty of time to produce your report once you have carried out the practical work. You are not under the same pressure as you are in a written test so you have the opportunity to submit a good piece of work which should enable you to get a high grade. Good luck!

Unit 3

Answers

Unit 1 Molecules and cells

Topic 1 Biological molecules

Testing your knowledge and understanding

1.1 $C_n(H_2O)_n$ or $C_nH_{2n}O_n$;

1.2 5C – ribose, ribulose; 6C – glucose, fructose, galactose;

1.3 5C = $C_5H_{10}O_5$; 6C = $C_6H_{12}O_6$;

1.4 6;

1.5 glucose; galactose; fructose;

1.6 respiring cells; blood plasma; gut;

1.7 respiratory substrates; supply energy;

1.8 maltose; lactose; sucrose;

1.9

Name	Location	Role
maltose	germinating barley	provides energy for developing embryo
sucrose	sugar cane	food source of pollinating insects / way in which carbohydrates synthesised in green plants transported through plant
lactose	in the milk of mammals	provides energy for developing offspring

1.10 maltose = glucose + glucose; sucrose = glucose + fructose; lactose = glucose + galactose;

1.11 $C_{12}H_{22}O_{11}$;

1.12

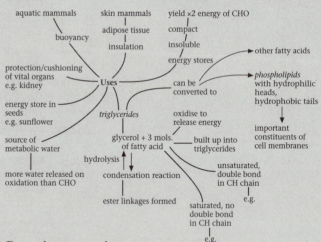

Key
In the mark schemes the following symbols are used:
; indicates separate marking points
/ indicates alternative marking points
eq. means correct equivalents points are accepted.

1.13

Name	Location	Sub-units Solubility	Roles in plants and animals
starch	plant cells	α-glucose; insoluble	energy storage as grains in plant cells
glycogen	animal cells; particularly liver and muscle; also bacteria	α-glucose; insoluble	energy storage in animal cells and bacteria
cellulose	plant cell walls	β-glucose; insoluble	structural component of the plant cell wall

1.14

aquatic mammals   skin mammals   yield ×2 energy of CHO

buoyancy   adipose tissue   compact

protection/cushioning of vital organs e.g. kidney   insulation   insoluble

energy store in seeds e.g. sunflower   Uses   energy stores   other fatty acids

triglycerides   can be converted to   *phospholipids* with hydrophilic heads, hydrophobic tails

source of metabolic water   oxidise to release energy   important constituents of cell membranes

glycerol + 3 mols. of fatty acid   built up into triglycerides

hydrolysis   unsaturated, double bond in CH chain e.g.

more water released on oxidation than CHO   condensation reaction

ester linkages formed   saturated, no double bond in CH chain e.g.

Practice questions

1

Statement	Triglyceride	Glycogen
contains only carbon, hydrogen and oxygen	✔	✔
glycosidic bonds present	✗	✔
soluble in water	✗	✗
provides storage of energy	✔	✔
occurs in flowering plants and animals	✔	✗

(Total 5 marks)

2 (a) (i) both amino acids drawn correctly; removal of water shown correctly; correct structure of dipeptide bond; **[3]**

(ii) condensation / polymerisation; **[1]**

(b) bonding / interactions within molecule / S–S bonds / H–H bonds; forming active sites of enzymes; reference to tertiary structure; form receptors in membranes; **[2]**

(Total 6 marks)

3 (a) A = deoxyribose; B = phosphate group; C = organic base / named example; D = hydrogen bond; **[4]**

(b) ring should be drawn around a phosphate + deoxyribose + base such as top left-hand corner; **[1]**

(c) (i) diagram should show double strand opening up; free nucleotides attaching to each strand; correct base pairing; **[3]**

(ii) each newly formed double helix contains one of the original polynucleotide strands; and one new polynucleotide strand; **[2]**

(d) S phase; **[1]**

(Total 11 marks)

4 correct details of bond formation; the nature of the glucose monomers in starch, glycogen and cellulose; reference to the different forms of branching in the polysaccharides; starch and glycogen as storage molecules; starch in plants, glycogen in animals; characteristics of starch and glycogen which make them good storage molecules / references to insolubility or osmotic effects; both can be broken down to release energy; cellulose in plant cell walls; structural support; description of the molecule / some mention of bonding between chains; formation of microfibrils;

(Total 10 marks)

Topic 2 Enzymes

Testing your knowledge and understanding

2.1 proteins; catalysts; soluble; low concentrations because of fast turnover rates; specific; lower activation energy;

2.2 E + S $\rightarrow$ E–S $\rightarrow$ E–P $\rightarrow$ E + P;

2.3 enzyme concentration; substrate concentration; temperature; pH; type and concentration of inhibitors;

2.4 Active site: part of enzyme, complementary to substrate, where substrate binds to R groups and reaction occurs;

Maximum rate: fastest rate of reaction at which the substrate is used or the product appears;

Optimum pH: the pH at which the maximum rate is achieved;

Turnover number: the number of substrate molecules that one molecule of enzyme can convert to product in one minute;

2.5 (a) denaturation;

(b) active site-directed inhibition;

(c) non-active site-directed inhibition;

2.6

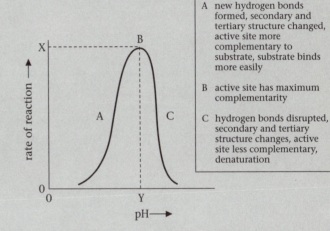

A new hydrogen bonds formed, secondary and tertiary structure changed, active site more complementary to substrate, substrate binds more easily

B active site has maximum complementarity

C hydrogen bonds disrupted, secondary and tertiary structure changes, active site less complementary, denaturation

2.7

Enzyme	Substrate	Product	Other
protease	protein	peptides, amino acids	used in biological detergents
pectinase	pectin	galactose, arabinose	clarifying fruit juice
lactase	lactose	glucose + galactose	can be immobilised for use in treatment of milk
sucrase	sucrose	glucose + fructose	present in succus entericus (intestinal juice)
maltase	maltose	glucose + glucose	present in succus entericus (intestinal juice)
amylase	amylose (starch)	maltose	present in saliva, pancreatic juice

Practice questions

1

Enzyme	Substrate	Type of reaction	Product	Commercial use
pectinase	pectin	*hydrolysis*	mono-saccharides disaccharides galacturonic acid	*clarifying fruit juice*
lactase	*lactose*	hydrolysis	*glucose and galactose*	*lactose free milk*
DNA polymerase	*mono-nucleotides*	*condensation*	*DNA*	DNA fingerprinting

(Total 8 marks)

2 (a) enzyme trapped in medium; medium is permeable; immobilised enzyme can be reused; **[2]**

(b) change in shape of the active site; due to change in bonding; substrate no longer fits; **[2]**

(c) energy required to start reaction; supplied by heat / chemical reaction; less required in presence of enzyme; **[2]**

(Total 6 marks)

Topic 3 Cellular organisation

Testing your knowledge and understanding

3.1 150 000 ÷ 10.5 = 14285.7;

3.2 **(a)**

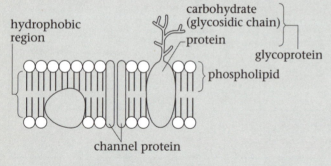

channel protein

(b) ribosome; microtubule; centriole; nucleolus;

(c) smooth and rough endoplasmic reticulum; Golgi apparatus; lysosome;

(d) mitochondrion; chloroplast; nucleus;

3.3 (a) nucleus and nucleolus (RNA, DNA); chloroplast and mitochondria (RNA, DNA); rough endoplasmic reticulum (RNA); ribosome (RNA);

(b)

Similarity	Chloroplast	Mitochondrion
Membranes	double	double
Electron transport	lamellae	inner membrane
Enzyme systems	Calvin cycle in stroma	Krebs cycle in matrix
DNA	in stroma	in matrix
RNA	ribosomes	ribosomes

3.4 freely permeable membrane; concentration gradient;

3.5 selectively permeable membrane; water potential gradient;

3.6 facilitated diffusion acts in the direction of a concentration gradient; active transport acts against it; facilitated diffusion does not require an energy source, active transport does and this is often ATP;

3.7 cell–cell interaction; receptor; cell-surface recognition; orientation of glycoproteins;

3.8 centrioles and microtubules make up the spindle to which the chromosomes from the nucleus become attached; the fibres (made of microtubules) contract;

3.9

Tissue	Function
upper epidermis	secretes cuticle; limits water loss
palisade mesophyll	absorption of light; photosynthesis
spongy mesophyll	some photosynthesis; region of large air spaces for gas movement
lower epidermis	site of stomata involved in water loss and gas entry and exit

Practice question

1

Organelle	Phospholipid	DNA	RNA
ribosome	✔	✗	✔
chloroplast	✔	✔	✔
Golgi apparatus	✔	✗	✗
smooth endo-plasmic reticulum	✔	✗	✗
mitochondrion	✔	✔	✔

(Total 5 marks)

Topic 4 The cell cycle

Testing your knowledge and understanding

4.1 (a) B(G$_1$) because both structural proteins and enzymes are being synthesised;

(b) C(S);

(c) D(G$_2$);

(d) A;

(e) B;

(f) D, to provide energy reserves for the process of photosynthesis;

4.2 (a) disruption of the hydrogen bonds holding the bases together;

(b) in the nucleus of the cell;

(c) each new double helix has one original and one new strand;

(d) catalyses the polymerisation in the 5' to 3' direction;

(e) joins the 3' to 5' sections of DNA together;

4.3 prophase (P) – **(g)** / 1, **(c)** / 2, **(d)** / 3, **(h)** / 4, **(n)** / 5, **(o)** / 6, **(a)** / 7;

metaphase (M) – **(p)** / 1, **(k)** / 2;

anaphase (A) – **(j)** / 1, **(b)** / 2, **(m)** / 3;

telophase (T) – **(i)** / 1, **(l)** / 2, **(f)** / 3, **(e)** / 4;

Practice questions

1 interphase; chromatids; centromere; nucleolus; nuclear membrane; metaphase; anaphase; pole; telophase; cell plate / phragmoplast;

(Total 10 marks)

Examiner's comments

It is best to read the passage first, before trying to put any words in the spaces. Then think carefully about what you should put in, so that the passage makes sense.

2 (a) quantity of DNA doubles; replication of the chromosomes / DNA; in preparation for mitosis / nuclear division; **[2]**

(b) period of rapid growth of the cell; synthesis of new organelles; protein synthesis; **[2]**

(c) cell organelles divide; accumulation of energy stores; chromosomes begin to condense; **[3]**

(d) DNA content halves / returns to original level; chromatids shared between daughter nuclei; **[2]**

(Total 9 marks)

Unit 2B Exchange, transport and reproduction

Topic 1B Exchanges with the environment

Testing your knowledge and understanding

1B.1 thin; moist; large surface area; transport of gas to surface; transport of gas away from surface;

1B.2 upper epidermis; palisade mesophyll; spongy mesophyll; lower epidermis; guard cells;

1B.3 ATP synthesised by chloroplast; H$^+$ ions pumped out of guard cell; K$^+$ ions enter; water potential decreases; water enters by osmosis; cells become more turgid;

1B.4 nose; larynx; trachea; bronchus; bronchiole; alveolus; capillary wall;

1B.5 external intercostal muscles contract; internal intercostal muscles relax; muscle of diaphragm contracts; diaphragm pulled flatter; thoracic volume increases; pressure decreases; air inhaled; lungs inflate;

1B.6 known mass of organisms; equilibration; screw clip closed; record time for coloured liquid to travel set distance; repeat to obtain mean;

1B.7

1. ribs	attached to sternum; attachment of intercostal muscles; physical protection; changes volume of thorax and therefore internal pressure
2. intercostal muscles	when external contract the ribs are pulled up; when internal contract the ribs are pulled down
3. lungs	bronchial tree supports airway; ending in alveoli for gas exchange; elastic so can expand and return to original size
4. diaphragm	central part fibrous; outer part muscular so can contract and become flatter; contributes to change in thoracic volume
5. pleural membranes	create airtight cavity around lungs; secrete fluid which acts as lubricant

Examiner's comments

A third column that contains details of other structures and connections could be added to this.

1B.8 mouth; oesophagus; stomach; duodenum; ileum; colon; rectum; anus (gall bladder and pancreas release secretions used in the duodenum);

1B.9 salivary amylase; pancreatic amylase; maltase; sucrase; lactase;

1B.10

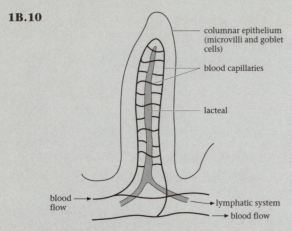

columnar epithelium (microvilli and goblet cells)

blood capillaries

lacteal

blood flow

lymphatic system

blood flow

Practice question

1 **(a)** A = squamous epithelium / alveolus; B = endothelial cell / capillary; **[2]**

(b) both have cells that are (very) thin / flat / eq.; A and B are in contact / eq.; so gases do not have to diffuse over long distances / eq.; any reference to cell surfaces being moist / reference to surfactants; **[3]**

(c) lowers the surface tension of the fluid on the alveolar surface; ensures that alveoli do not collapse; **[2]**

(Total 7 marks)

Topic 2B Transport systems

Testing your knowledge and understanding

2B.1(a) 6:1;

(b) 3:1;

(c) it gets smaller, so as the bulk of an organism increases there is less surface area available for the exchange of materials;

2B.2 vessels; tracheids; fibres; xylem parenchyma;

2B.3 sieve tube elements; companion cells;

2B.4 wind / air movements – increase in wind increases transpiration; temperature – increase in temperature increases transpiration; light intensity – increase in light intensity increases transpiration; humidity – increase in humidity decreases transpiration;

2B.5

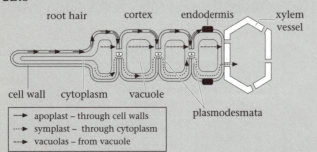

root hair   cortex   endodermis   xylem vessel

cell wall   cytoplasm   vacuole

plasmodesmata

→ apoplast – through cell walls
····▶ symplast – through cytoplasm
---▶ vacuolas – from vacuole

2B.6 transport of respiratory gases (carbon dioxide, oxygen); transport of nutrients (e.g. glucose, minerals, amino acids); transport of metabolic waste (carbon dioxide, urea); transport of hormones; defence against disease (clotting, antibodies);

Examiner's comments

When you are asked for two or three functions choose clearly different functions. You should not give transport of glucose and transport of amino acids as they are both nutrients.

2B.7 1. aorta / aortic arch;
2. pulmonary artery;
3. pulmonary vein;
4. left ventricle;
5. coronary arteries;
6. posterior vena cava;
7. right ventricle;
8. right atrium;
9. anterior vena cava;

2B.8 vena cava, right atrium, right ventricle, pulmonary artery, lung capillaries, pulmonary vein, left atrium, left ventricle, aorta;

2B.9 structural differences – tunica media thinner (less smooth muscle, fewer elastic fibres); presence of semi-lunar valves; blood differences – deoxygenated; lower pressure; slower flow; flowing towards the heart;

2B.10 biconcave – gives large surface area to volume ratio, shorter distance for diffusion; haemoglobin carries oxygen as oxyhaemoglobin and carbon dioxide as carbaminohaemoglobin, small decrease in pH can cause large release of oxygen from oxyhaemoglobin;

2B.11

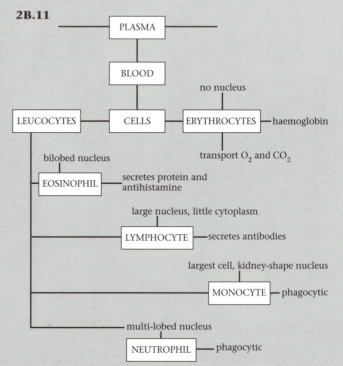

PLASMA

BLOOD

no nucleus

LEUCOCYTES   CELLS   ERYTHROCYTES — haemoglobin

transport O_2 and CO_2

bilobed nucleus

EOSINOPHIL — secretes protein and antihistamine

large nucleus, little cytoplasm

LYMPHOCYTE — secretes antibodies

largest cell, kidney-shape nucleus

MONOCYTE — phagocytic

multi-lobed nucleus

NEUTROPHIL — phagocytic

2B.12

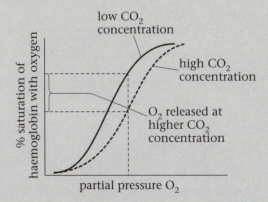

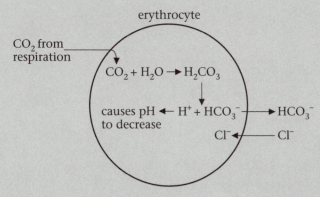

2B.13

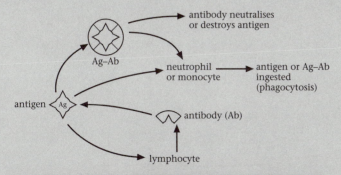

Practice questions

1

Cell type	*One* **characteristic structural feature**	*One* **function**
sieve tube element	*sieve plate / no nucleus / few organelles;*	*translocation / transport of sucrose / amino acids;*
vessel / tracheid;	*lignified wall / no cell contents / no end wall;*	transport of water and mineral ions
fibre;	narrow lumen	support

(Total 5 marks)

2 **(a)** read figures accurately from graph: 61 ±1 and 175 ±1; [(175 − 61) ÷ 61] × 100; 186.8 %; **[3]**

(b) overall increase in transpiration as pore size increases; steep increase up to 7.5 / greatest rate of change; above 7.5 increase in pore diameter has less effect; **[2]**

(c) transpiration rate greater / eq.; continues increasing / linear relationship / rate does not decrease; **[2]**

(d) water vapour removed from vicinity of leaf in moving air / converse for still air; steeper concentration gradient / converse for still air; so more water vapour lost / faster rate of transpiration / converse for still air; the wider the stomatal aperture the more water vapour is lost; **[3]**

(e) temperature – increase will increase rate / eq.; humidity – increase will decrease rate; light intensity – will affect stomatal aperture; **[2]**

(Total 12 marks)

3 **(a)** bone marrow; **[1]**

(b) granulocyte / granular polymorph / neutrophil / eosinophil;
monocyte; **[2]**

(c) platelet / thrombocyte; **[1]**

(d) to enable continued formation of blood cells / eq.; **[1]**

(Total 5 marks)

Topic 3B Adaptations to the environment

Testing your knowledge and understanding

3B.1 Xerophyte: plant that lives in low water areas, e.g. sand dunes, able to survive drought;

Hydrophyte: plant that lives in an aquatic or very wet habitat, adaptations to water movement and depth;

3B.2 high temperature; low humidity; increased air movement; sunlight;

3B.3 the epidermis; spiracles and tracheoles; trapped air bubble; gills; breathing tube;

3B.4

Structure	Organism	Source of oxygen	Efficiency due to
epidermis	planarian	dissolved	diffusion gradient and large surface area to volume ratio
gills	*Daphnia* sp.	dissolved	diffusion gradient; large surface area; thin
air bubble / spiracle	diving beetle	air	spiracles in contact with air; diffusion gradient
breathing tube	mosquito larva	air	spiracles in contact with air; diffusion gradient
stored haemoglobin	*Tubifex*	dissolved	combines with oxygen; diffusion gradient

Practice question

1 **(a)** A= breathing tube;   **[1]**

   (b) tube can reach surface / extensible; breathing tube connected to tracheal system; ventilates using air from the surface;   **[3]**

   (c) diffusion; through epidermis;   **[2]**

                             (Total 6 marks)

Topic 4B Sexual reproduction

Testing your knowledge and understanding

4B.1 seminiferous tubules; mitosis; primary spermatocytes; meiosis; spermatids; secondary spermatocytes; spermatozoa; oogonia; diploid; primary oocytes; meiotic; secondary oocyte; polar body; fertilisation;

4B.2 I – **(a)**, **(c)**, **(d)**, **(f)**, **(i)**, **(k)**, **(l)**;
W – **(b)**, **(e)**, **(g)**, **(h)**, **(j)**;

4B.3(a) dioecious;

   (b) endosperm;

   (c) protogynous / protogyny;

   (d) cross-pollination;

   (e) protandry / protandrous;

4B.4 (a) A = scrotum; B = testis; C = epididymis; D = seminal vesicle; E = vas deferens; F = prostate gland; G =urethra; H = vagina; I = cervix; J = ovary; K = oviduct; L = uterus;

   (b) (i) H; **(ii)** B; **(iii)** J; **(iv)** D / F; **(v)** C; **(vi)** J; **(vii)** G; **(viii)** K;

4B.5

Hormone	Source of secretion	Stimulus for secretion	Effects
follicle stimulating hormone	anterior pituitary	fall in progesterone levels (-ve feedback); increasing levels of oestrogen (+ve feedback);	stimulates development of primary follicle; stimulates secretion of hormones from ovary (oestrogen and progesterone)
luteinising hormone	anterior pituitary	fall in progesterone levels; increase in oestrogen levels (as above)	triggers ovulation; conversion of follicle to corpus luteum; progesterone secretion
oestrogen	ovary; after ovulation from corpus luteum	increasing FSH levels	responsible for development of secondary sexual characteristics; repair of endometrium after menstruation; increase in FSH during follicular phase
progesterone	ovary; corpus luteum after ovulation	increasing FSH levels; secretion of luteinising hormone	high levels inhibit LH; maintenance of the endometrium in secretory phase of menstrual cycle
prolactin	anterior pituitary	suckling	stimulates synthesis and secretion of milk
oxytocin	posterior pituitary	suckling	ejection of milk into ducts; contraction of uterine muscle during labour

Note: Not all the hormones are included. The prostaglandins and their effects have been omitted. You could include these to make the table more complete.

Practice questions

1   **(a)** meiosis;   [1]

   (b) (i) anther / ovule / pollen sac / microsporangium / megasporangium;   [1]

      (ii) ovary / testis / seminiferous tubule;   [1]

   (c) halves chromosome number / produces haploid cells / haploid gametes; reference to the introduction of genetic variation;   [2]

(Total 5 marks)

Examiner's comments

A very straightforward question, but think very carefully about the division as it has a knock-on effect.

2   follicle-stimulating hormone; primary follicles; oestrogen; endometrium / lining of uterus; secondary oocyte; corpus luteum; progesterone;

(Total 7 marks)

Examiner's comments

Alternatives to the names of some of the hormones are acceptable. For example, you might be more familiar with the term 'follitropin' for 'follicle stimulating hormone' – FSH would also be acceptable. 'Yellow body' would be acceptable for the 'corpus luteum'.

Unit 2H Exchange, transport and reproduction in humans

Topic 1H Exchanges with the environment

Testing your knowledge and understanding

1H.1 1. squamous, alveolus; 2. cuboidal, nephron; 3. columnar, ileum;

1H.2 **A** resting state / normal breathing rate; **B** increasing physical activity, muscular contraction; **C** tidal volume / vital capacity, need for more oxygen / elimination of more carbon dioxide; **D** residual volume, volume of air in alveoli that cannot be expired;

1H.3

Smoke content	Effect on ventilation	Effect on gas exchange	Effect on pregnancy
Carbon monoxide	Increases ventilation rate	Reduces uptake of oxygen due to the formation of carboxy-haemoglobin	Hypoxia reduces availability of oxygen to fetus, reduces rate of growth
Nicotine	Paralysis of cilia	Accumulation of mucus in bronchioles. Reduces gas exchange	N/A
Tars (hydro-carbons, phenols, fatty acids)	Increases mucus production	Tumors, emphysema, bronchitis reduce area for gas exchange	N/A

Practice question

See *Answers*, Unit 2B, Topic 1B.

   (c) (i) alveolar walls destroyed / lumen contains mucus / surface area for exchange of gases reduced in size / reference to loss of elasticity;   [1]

      (ii) slows down / decreases rate (of diffusion);   [1]

Topic 2H Transport of materials

Testing your knowledge and understanding

2H.1 generates electrical impulse on demand or when heart misses a beat; conducts to heart muscle; causes contraction of cardiac muscle; to replace or aid natural pacemaker;

2H.2 **P** wave, electrical excitation of the atria; **QRS** complex, excitation of the ventricles; **T** wave, repolarisation of the ventricles;

Practice question

1   **(a)** consist of a single layer of cells / simple epithelium / endothelium; cells very thin / flattened / squamous; presence of pores; selectively / partially permeable/permeable to small molecules;   [2]

   (b) drains into lymphatic vessels / lymphatic capillaries / lacteals; to form lymph; moves in lymphatic vessels by local body movement; reference to valves / one way flow in lymphatic vessels; lymph returned to venous system;   [3]

   (c) high blood pressure / decrease in plasma proteins / protein deficient diet / increased capillary permeability / blocked lymphatic vessels;   [1]

(Total 6 marks)

Topic 3H Human ecology

Testing your knowledge and understanding

3H.1 diurnal (temperature variation); early morning, e.g. ~02.00 to 04.00 hrs; ~1.5 °C (can vary in different people); hypothermia sets in at ~35 °C but body can go lower for limited periods (e.g. use in medical operations or occasional amazing survival stories – can you find details of any? – e.g. report in *Independent*, 28 January 2000 [full report by Professor Gilbert in *The Lancet*] describing survival of woman trapped under ice for 40 minutes and her body temperature dropped to 13.7 °C); 42 °C;

3H.2 (a) A = *convection*; B = *molecules*; *warm*; *cooler*; C = layers of loose clothing, designed to trap air close to skin – lightweight rather than heavy to reduce energy expenditure during physical exertion;

(b) A = *conduction*; B = *contact*; C = padding, layers, external 'waterproof' layer, particularly need to insulate extremities (fingers, toes);

(c) A = *radiation*; B = *gain*; *reflection*; *loss*; C = need to protect bare skin such as face, fingers from loss of heat and damaging effects of radiation;

(d) A = *evaporation*; B = *evaporation*; *vaporisation*; *cooling*; C = as for convection (to reduce loss of heat) – but note conflict between conserving heat and the body becoming uncomfortable inside waterproof clothing because of increasing humidity as moisture (from sweating) accumulates – ideally clothing should have some permeability to allow water loss;

3H.3 (a) *clothing* – loose, perhaps more than one layer (but note that while this helps to avoid excessive loss of water there is conflict with sweating as a cooling mechanism if water cannot evaporate); light colours to reflect radiation;

buildings – thick walls; made of material which insulates well; small windows; system for ventilation (fans, air inlets etc.); white to reflect; shade for protection from direct radiation;

(b) in *clothing* – emphasise need for sweat to evaporate from body;

in *buildings* – more emphasis on ventilation and movement of air;

3H.4

Cold injury	damage to tissues due to low temperatures; reduced blood supply to tissues means reduced supply of nutrients and oxygen, interferes with normal metabolic activities;
Frostbite	tissues freeze, mechanical action of ice crystals cause damage to cell structure, liquid water withdrawn leading to dehydration; if superficial, tissues can recover, but damage may be permanent;
Trench foot	from cooling in cold water; tissues blackened, damage to muscles and nerves, can recover if warmed up, but if not, can lead to gangrene;
Hypothermia	core temperature below 35 °C; reduction of heart rate; muscle weakness, then – as temperature falls further – loss of coordination, mental confusion, loss of consciousness;

3H.5

Sunburn	damage to skin with reddening and blistering, due to excessive radiation;
Prickly heat	irritation of skin, ducts from sweat glands blocked;
Heat collapse	dizziness; blood diverted to skin, away from internal organs;
Heat exhaustion	loss of body fluid leading to dehydration if liquid not replaced; loss of body salts (particularly sodium chloride); low blood volume and low blood pressure; fatigue, followed by mental and physical deterioration; further loss means fluid drawn out of cells leading to cell damage;
Heat stroke	breakdown of temperature regulation mechanisms; irreversible damage to cells and proteins as body core temperature exceeds 42 °C or above;

3H.6 (a) look back to your answers for question 3H.2 (and remember to think about metabolic heat production, hot food and drinks);

(b) you can work out for yourself;

(c) temperature receptors in skin;

(d) detectors in hypothalamus;

(e) hypothalamus;

3H.7 *temperature receptor* – nerve endings, sensitive to changes in temperature in environment;

sweat glands – when hot, secrete more liquid (known as sweat), sweat duct carries liquid to skin surface, liquid / sweat evaporates and causes cooling (*note*: evaporation slow or non-existent in highly humid conditions);

hair erector muscle – when cold, contracts and raises hairs; hairs trap layer of air which acts as insulator (effect relatively small in humans); when hot, hairs lie flat allowing more air movement, hence cooling of skin surface;

adipose tissue – structural feature, provides insulation from cold (with longer term benefit);

arterioles, blood capillaries, capillary loop – in cold conditions – arterioles constrict, diverts blood away from skin, reduces blood supply to capillary loops, reduces blood supply near skin surface, reduces heat loss; in hot conditions – arterioles dilate, allows more liquid to be taken up by sweat glands and more blood near skin surface, hence helps with cooling;

3H.8 (a) *adaptations of natives in hot climates* – tall thin physique and large body surface area (to volume ratio); high rate of sweat loss; traditional suitable clothing;
adaptations of natives in cold climates – short fatter physique with smaller body surface area (to volume ratio); Inuits have higher blood flow through extremities; shape of nose in relation to loss of heat in ventilation; traditional suitable clothing;

(b) *visitors to hot climates* – acclimatisation mainly as increased sweat loss; sweat produced has lower salt concentration; adopt suitable clothing;
visitors to cold climates – acclimatisation probably limited; may have increased appetite; hence increased food intake; otherwise mainly behavioural, e.g. adopt suitable clothing, voluntary increase in physical activity;

3H.9 *Thermogenesis*: heat produced in the body from metabolic activities (e.g. in liver and other organs and in skeletal muscle);
Thermoregulation: temperature control within narrow range; homeostatic mechanism;

Vasoconstriction: constriction of arterioles so that the lumen narrows, e.g. in heat conservation, occurs below skin surface and diverts blood away from capillaries in skin, hence conserves heat in body;
Vasodilation: dilation or opening of lumen of arterioles, e.g. in cooling mechanism, occurs below skin surface allowing more blood to flow through capillaries in skin, hence increases loss of heat from body;

3H.10 Himalayas (e.g. people in Nepal, Bhutan, China including Tibet and other mountain areas); Peruvian Andes (Quechua Indians etc.);

3H.11

Environmental condition	Stresses and effects on the body
low temperature	cold; leading to hypothermia / frostbite;
low humidity	high rate of evaporation; loss of water from breathing and through skin surface, leading to dehydration; skin becomes dry, leading, for example, to cracked lips;
high winds	increases rate of evaporation of water, leading to dehydration (see low humidity); increases wind chill effect, hence effects of cold (hypothermia / frostbite)
increased solar radiation	high daytime temperatures (increased by reflection from snow), leading to loss of water (sweating); high UV radiation, leading to skin damage / damage to cornea of eye (snow blindness);

3H.12 (see comments in question 3H.2 on clothing under cold temperature extremes) – remember the importance of clothing being windproof and the dangers of the 'wind chill' factor, but be aware of the conflict between protection from wind and discomfort experienced with increased humidity within clothing layers from sweating; remember need to protect eyes with goggles or dark glasses, and exposed skin from radiation and from cold ;

3H.13 (a) (i) A = *hyperventilation*; B = *increased; more; deeper; increased*; C = *higher intake of oxygen*; D = *higher output of carbon dioxide; alkaemia*;

(ii) A = *increased pulmonary diffusing capacity*; B = *increased surface area of lungs; increase in blood flow*; C = *increased diffusion of oxygen, alveoli*;

(b) (i) A = *increased cardiac output*; B = *increased heart rate; increased stroke volume*; C = *more blood pumped; increased collection of oxygen*;

(ii) A = *increased haemoglobin*; B = *increased number of red blood cells; increase in haemoglobin concentration*; C = *increased carrying capacity*; D = *reduce blood flow*;

(iii) A = *oxygen–haemoglobin dissociation*; B = *shifts to right*; C = *released more easily*;

(c) (i) A = *decreased ADH secretion*; B = *more urine*;

(ii) A = *increased ADH secretion*; B = *less urine*; C = *excess; oedema; pulmonary oedema; cerebral oedema*;

3H.14 not clear cut, though native highlanders are often robust and able to undertake heavy physical work at high altitudes – some show increased lung volume + barrel-shaped chest; hyperventilation;

3H.15 increased heart rate (1); increased breathing rate (2); increased concentration of haemoglobin in the blood (3); increased red blood cell production (4); increased capillary density (5);

3H.16 (a) headaches; lack of concentration; disinclination to work; giddiness; insomnia; breathlessness; coughing and difficulty with breathing; increased heart rate; palpitations; nausea; vomiting, anorexia; oedema of legs and feet;

(b) get down to lower altitude as quickly as possible;

Practice question

1 (a) temperature at centre of body / deep in body / eq.; kept within narrow limits; example of temperature range (*accept within range 36 °C – 38 °C*);   **[2]**

(b) reduce number of variables / eq. / female percentage fat higher and could affect heat loss / eq. / menstrual cycle ref. ovulation;   **[1]**

(c) (i) European drops more quickly in first hour / only drops in first hour; European remains more constant / changes less (after 1.5 hours); aborigine has steadier fall for 8 hours / eq.; European falls by approximately 1 °C, aborigine falls by more than 2 °C / falls × 2;   **[3]**

(ii) aborigines tolerate lower core temperature during cold nights / eq.; body heats up in morning as air temperatures rise; aborigines have less need of cooling mechanism in morning as core temperature increases / eq.;   **[2]**

(d) increased metabolic rate / thermogenesis; exercise / shivering; hair erection / vasoconstriction;   **[2]**

(e) hypothermia; reduced cardiac / heart rate / output; reduced blood flow to coronary / cerebral circulation; increased urine output / cold diuresis; difficulty walking / coordinating movement / fatigue; loss of consciousness / mental confusion / disturbed vision;   **[3]**

(Total 13 marks)

Topic 4H Human reproduction and development

Testing your knowledge and understanding
See *Answers*, Unit 2B, Topic 4B

Practice question

1 (a) axes correctly labelled with units and right way round; adequate scale used; plotting of all points correct for each hormone; points joined correctly with a straight line; key to the curves;   **[5]**

Examiner's comments

Plotting graphs is considered a skill and you are unlikely to have to do this on a theory paper very often, but it is as well to be prepared. Make sure you get the axes the right way round and take care with the positioning of the points and the drawing of the curves. The rest of the question makes you use the information on the graph and interpret it using your biological knowledge.

(b) no HCG for two weeks; rapid rise until 8 weeks; starts with implantation; produced by placenta / chorion / trophoblast / eq.; to maintain corpus luteum / maintain progesterone level / eq.; until progesterone produced by placenta;   **[4]**

(c) development of endometrium / eq.; stimulates growth of blood supply to placenta; inhibits FSH production; inhibits LH production / ovulation; inhibits oxytocin production; mammary gland development during pregnancy;   **[3]**

(d) development of baby requires increasing supplies / eq.; uterus / placenta / blood supply must be kept up; quantity needed for FSH / LH inhibition unlikely to change; oxytocin inhibition may need to increase; mammary gland development increases towards end of pregnancy;   **[2]**

(Total 14 marks)

Unit 3 Energy and the environment

Topic 1 Modes of nutrition

Testing your knowledge and understanding

1.1 (a) holozoic / any named animal;

(b) autrophic / green plants / any named green plant;

(c) saprobiontic / saprophytic / any fungus / *Rhizopus*;

(d) herbivore / cow / ruminant;

(e) parasite / *Taenia*;

(f) mutualistic / symbiotic / *Rhizobium* in pea / bean plants;

(g) carnivore / cat / dog / lion / any named carnivore;

1.2 (a) fungi;

(b) platyhelminths;

(c) bacteria;

1.3 herbivore – **(b)**, **(d)** **(e)**, **(f)**, **(h)**; carnivore – **(a)**, **(c)**, **(g)**;

1.4

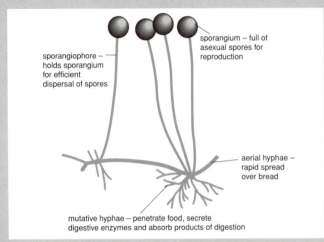

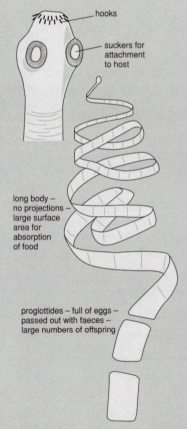

1.5

Practice question

1 abomasum – digestion of proteins / true stomach / produces enzymes / acid;

omasum – removes water / liquid / organic acids from food;

reticulum – allows passage of food to and from rumen and oesophagus / formation of balls of cud;

rumen – fermentation chamber / food churned / bacterial digestion;

(Total 4 marks)

Topic 2, 3 Ecosystems and energy flow

Testing your knowledge and understanding

2,3.1 temperature; oxygen concentration; sunlight (hours / wavelength); soil (type / pH); water (availability / pH / salinity); air / water movements;

2,3.2 Biosphere: part of Earth and atmosphere occupied by organisms;

Habitat: place where an organism lives;

Consumer: cannot synthesise own organic food so needs to obtain it from other organisms;

2,3.3

$$\text{carbon dioxide} + \text{water} \xrightarrow[\text{chlorophyll}]{\text{sunlight}} \text{glucose} + \text{oxygen}$$

2,3.4

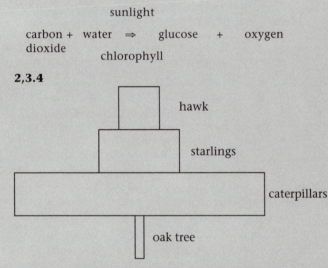

Organism	Contribution to relationship	Benefit from relationship
Rhizobium in root nodules of leguminous plant	fixes atmospheric nitrogen into ammonium compounds which plant can use in the synthesis of amino acids and proteins; enables plant to thrive in nitrogen-deficient soils	gains sugars and other organic compounds (products of photosynthesis) from plant
microorganisms in the rumen of a cow (ruminant)	some produce cellulose-digesting enzymes (cellulases) which break down the plant cell walls; cellulose broken down to hexoses which are fermented by other microbes to organic acids under anaerobic conditions; some protein also broken down to ammonia; some B-complex vitamins also synthesised	the different microbes present all benefit from the availability of a food source and an equable temperature, i.e. a favourable habitat in which to grow and reproduce

2,3.5 inedible material; undigested material (faeces); heat from respiration / metabolism; excretory material; dead organisms;

Examiner's comments

Take care when answering this type of question. Answers such as 'movement' and 'growth' are a little too vague. It is better to use the term *metabolism*. Remember that, although the material or energy is lost from each level, it may be available to decomposers.

2,3.6

Producers	Primary consumers	Secondary consumers
higher plants	mammals	mammals
algae	birds	birds
	bacteria and fungi	fish
	arthropods	arthropods
	molluscs	microfauna
	annelids	
	microfauna	

2,3.7 collect a water sample of 1 m³; collect several samples from different areas to obtain mean; identify organisms and sort into trophic levels; weigh all the organisms in each tropic level; decide what area on the diagram each 1 cm³ will represent; draw a block of the correct area, to represent the producers; complete the pyramid by adding the other blocks so that it is symmetrical;

Practice questions

1 (a) food chain is one chain of feeding relationships, food web shows all the food chains; organism in food chain occupies only one trophic level, organisms in food web can occupy more than one trophic level; **[2]**

(b) producer is autotrophic, decomposer is heterotrophic; producer's primary energy source is sunlight, decomposer obtains energy from organic source; **[2]**

(c) community is several populations, ecosystem has biotic and abiotic components; community populations interact, ecosystem has interactions between biotic and abiotic; **[2]**

(Total 6 marks)

2 (a) (i) primary consumers / herbivores;

(ii) secondary consumers / carnivores; **[2]**

(b) (i) $[(2.8 \times 10^4) \div (3 \times 10^6)] \times 100$; 0.9 %; **[2]**

(ii) some light reflected; some wavelengths not absorbed; lost as heat; another factor is limiting, e.g. temperature or carbon dioxide concentration; **[2]**

(Total 6 marks)

Topic 4 Recycling of nutrients

Testing your knowledge and understanding

4.1 transpiration; evaporation (evapotranspiration); precipitation; condensation;

4.2 respiration – converts carbohydrates to carbon dioxide;
photosynthesis – fixes carbon dioxide as carbohydrates;
carbon sinks – long-term forms of carbon and carbonates in bones, exoskeletons and limestones, as fossil fuels;

4.3

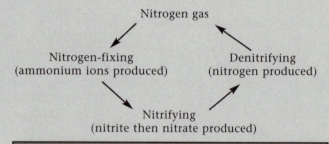

Examiner's comments

You could produce simple cycles that show the other organisms involved and the inorganic processes of lightning and use of fertilisers. Then put them all together for a complete picture.

4.4

Letter	Name of process	Conditions needed
A	*photosynthesis*	*sunlight, carbon dioxide*
B	*respiration*	*oxygen / aerobic*
C	*fossilisation*	*anaerobic*
D	*combustion*	*oxygen*
E	*methanogenesis*	*anaerobic*

4.5

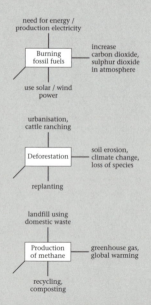

Practice question

1 (a) A = nitrogen fixation; B = nitrification; C = decomposition / putrefaction / ammonification; **[3]**

(b) (nitrate used for) increased growth of algae; (increased rate) death of algae; used as food source for bacteria; increased rate of growth / metabolism of bacteria; increased biochemical oxygen demand / BOD; reduction in oxygen availability for other organisms; **[4]**

(Total 7 marks)

Examiner's comments

This is the usual course of events and this account will gain you all the marks available. If you have started with a reduction in light penetration caused by the algal growth then you would gain marks for referring to the death of submerged aquatic plants and continuing the story from there.

Topic 5 Energy resources

Testing your knowledge and understanding

5.1 (a) oil; coal; natural gas;

(b) *coal* – decomposition former vegetation / organic material; accumulated in peat beds; waterlogged + anaerobic conditions; buried under sediments; compacted / compressed; becomes lithified / rock like;

oil – similar to coal, but found as liquid; usually in sedimentary rocks of marine origin;

natural gas – similar to oil and coal, but found as gas; in underground reservoirs in rocks; mainly methane, plus small amounts ethane, propane and other hydrocarbons;

(c) certainly wind and hydroelectric power; then, depending on your definition, probably also solar photovoltaics and geothermal power;

(d) reaction against dependence on fossil fuels because of pollution effects (such as acid rain); concerns over reaching limit of finite resources; looking for viable / economically acceptable alternative energy resources (without fossil fuel disadvantages); pressure / trend for using 'environmentally friendly' materials / NFFO or similar requirements;

5.2 (a) woody crops managed by short rotation coppice (e.g. willow, poplar and others); other densely planted wooded or forested areas (e.g. with conifers); *Miscanthus*; can also include sugar beet, oil-seed rape, sunflower, sorghum etc.;

(b) items from above list, but add sugar cane, cassava etc. – many species, temperate and tropical, could be exploited;

(c) suitable example – e.g. short rotation coppicing – choose woody species (e.g. willow); close planting distance; end first year cut to ground; then harvest by cutting after 2 / 4 etc. years' growth; wood chips used to generate electricity, or as appropriate for crop chosen;

(d) derived from renewable biomass; increase range of fuels available (as alternatives to fossil fuels);

5.3 (a) 80 % to 90 % unleaded petrol with 20 % to 10 % added ethanol; used as a motor fuel;

(b) often sugar crops (sugar cane, sugar beet); other crops include maize, manioc, potatoes, grain crops, even wood;

(c) anaerobic fermentation / respiration; usually by yeasts;

5.4 (a) 50 % to 75 % methane ('marsh gas') with carbon dioxide plus traces ammonia, hydrogen sulphide, water; used for fuel / heating / particularly small-scale or rural locations;

(b) organic materials, particularly waste including animal waste / slurries / manure / sewage;

(c) bacteria (for types, see stages); first aerobic then anaerobic digestion; several stages – hydrolysis, acetogenesis, methanogenesis;

5.5 (a) residues (after fermentation / digestion) used as fertiliser; disposes of waste (giving useful product); growing woody crops means more carbon dioxide used (gas balance); lower output of polluting gases;

disadvantages might include – unfavourable energy balance (i.e. amount of energy used to obtain a usable output of energy in a form it can be used); often very small-scale operations; insufficient for wider scale demand for energy; can be labour intensive, hence increases cost;

Examiner's comments

Remember to link your answers to the correct fuel. Not all of the answers given apply to all the fuels.

(b) answers here overlap with those you may have given to question 5.2 (see above), plus you might add straw or other waste surplus from harvest, dung (animal manure), coconut oil, sunflower oil and many other materials – see what you can find out

(c) here again, there are no simple answers, and it depends on which fuel you choose – but the question is partly to make you work out what you mean by 'sustainable' then consider as far as you can, whether this fuel source can be harvested continuously, and be replaced or replenished so that it does not run out, and without a high input of other resources (which also have an energy cost)

5.6 (a) derived from landfill sites / municipal wastes; mainly methane and carbon dioxide, though composition generally has lower proportion of methane (compared with biogas – see question 5.4 above); if not harnessed, leaks and escapes and could be dangerous and cause explosions, unless 'burnt off' at the site;

(b) probably from about 2 years up to 15 years (depends on the site);

(c) landfill gas (LFG) is perhaps not really 'renewable' as it is derived from waste materials, many of which have already been manufactured within human societies; but is a way of using a by-product from the disposal of waste and recovers some energy from it; methane, if allowed to escape, is a powerful greenhouse gas, so would make a greater contribution to the greenhouse effect than the carbon dioxide produced after combustion of methane; overall seems an advantageous way to harness energy from waste materials;

Practice question

1 (a) converts starch / amylose; to maltose; for respiration of carbon / carbohydrate source; eq.; **[2]**

(b) ethanol / eq.; petrol; **[1]**

(c) readily available / no imports needed; renewable energy source / petrol non-renewable / eq.; less air pollution / less sulphur content / less sulphur dioxide production; utilises waste products / cheaper; **[2]**

(d) yeast / animal feed / fertiliser / SCP [single-cell protein] / alcohol / carbon dioxide / eq.; **[1]**

(Total 6 marks)

Topic 6 Human influences on the environment

Testing your knowledge and understanding

6.1 obtain fuel, timber for building; land clearance – for other agricultural crops / increased grazing / urban development / industrial factories / roads through forest areas / etc.; other uses forest species – e.g. obtain medicines / etc.; 'slash and burn' shifting cultivation (but forest usually recovers);

6.2 long time, some examples to support this include – early Neolithic settlers, clearance for agriculture – (e.g. quote in Plato's writings 2300 years ago) – major forest clearance in late Bronze Age (e.g. in UK), medieval forest clearance (UK) – large-scale destruction of temperate forests in late nineteenth century – etc.;

6.3 *soil* – loss of vegetation, loses 'sponge-like' effect; increased surface run-off water, surface soil particles and nutrients washed into streams; leads to erosion; poorer quality soil if used for agriculture;

rivers – deposition of sediment; flooding;

nutrients – lower due to leaching (increased run-off of water); disturbance of microbes involved in recycling of nutrients from fallen leaves; surface organic litter etc.;

oxygen / carbon dioxide – replacement vegetation likely to have lower photosynthesis overall, hence net increase carbon dioxide with net decrease oxygen;

biodiversity – diversity decrease; could lead (has led) to loss of rare or endangered species;

6.4 *see question 6.3 (soil) above* – plus sudden exposure of soil to heat, becomes desiccated; loss of vegetation (tree) cover so more exposed to effects of wind and rain;

6.5 try to find some examples yourself – e.g. greater diversity in managed coppice woodland compared with surrounding cultivated farmland; giant panda and loss of habitat providing special food requirements; loss of large mammals in UK and European woodland (and finally aurochsen) associated with loss of forest cover; etc.;

6.6 crop of wood can be cut repeatedly; say 2 to 10 years (depending on coppicing cycle and purpose); ancient coppicing in UK still continues today (several hundred years);

6.7 most schemes involve controlled felling of trees in relation to controlled replanting, find specific examples if you can; another strategy is to set aside areas with high conservation value, e.g. as National Parks, and ensure this is managed (protecting indigenous people, protection to retain sufficient water etc.);

6.8 *deforestation*: permanent clearing of forest areas and subsequent utilisation of the land for a different non-forest purpose;

afforestation: planting trees on land that had previously not been used for forest;

reforestation (also *reafforestation*): planting of trees to replenish stock in forest or woodland areas from which trees have been removed (either by humans or natural causes); (*Note*: The term is used in connection with land that has been used for forest, say within living memory or previous 50 years.)

6.9 area already with low rainfall; supports limited vegetation cover; over-exploitation of that land by human activity, e.g. gathering fuel / overgrazing / growing crops; i.e. exceeds carrying capacity of the land; eroded soil particles may be deposited on other areas of marginal land (low agricultural economic return) making pressures on land greater;

6.10 *erosion* – lack of vegetation / plant cover; exposure of bare soil to sunlight; further desiccation; exposure to wind removes soil particles;

salinisation – high evaporation rate, low plant cover so salts drawn up and left on soil surface; becomes worse if chemicals, such as fertiliser or herbicides / pesticides used;

6.11 irrigation channels do not allow adequate flow of water; area tends to become waterlogged; salts brought up from groundwater; high evaporation rates means water, then soil surface, becomes much more salty; unsuitable for plant growth;

6.12 plant trees to create shade / shelter from wind; leguminous tree species give some nitrogen fixation (increases soil fertility); terraces on slopes to reduce run-off of water; furrows or depressions to help trap moisture / shelter for seedlings; stones on ground trap moisture / reduce evaporation; stones also give shade / shelter for seedlings / reduce evaporation around trees;

maintenance of area by controlled planting of crops or use as grazing land, but keep within productive capacity of the land – try to find actual examples to quote;

6.13 (a) 1. steep rise to ~1974, fluctuation approximately a plateau, slow rise again from 1988;
2. lower than industrial countries, but steady rise, steeper from ~1965;
3. steady rise to ~1988 (less than developing countries);

(b) *industrial* – 1950 to ~1980, steady increase in demand, then some reduction; perhaps less industrial expansion; improved technology brings increased energy efficiency (so burns less for same energy); use of alternative fuel sources; response to legislation and international agreements to reduce fossil fuel burning;

developing countries – increase in industrial demand (similar to an 'industrial revolution') particularly from early 1980s; expanding population with higher demands; higher domestic demands in response to social trends;

former Eastern Bloc – increased industrial output with increased demand; then noticeable decrease in late 1980s, linked to economic collapse because of political changes in the former Soviet Union;

Examiner's comments

Figures for actual concentrations are approximate, as read off from the graph. In an examination paper, you would usually be given a grid, so you can read off the figures more precisely. The table below gives stages you may take in working out your answer and shows different ways you can manipulate the figures. By doing this, it makes you look at the data and you become more aware of the trends that are taking place. Even if you are not asked to calculate anything, it is often worth 'probing' into the data a bit more to tease out useful and interesting information.

(c)

Year	1950	1960	1970	1980	1990	1997 (est.)
million tons carbon	1610	2520	3990	5160	5950	6390

[Note: 1 ton is approximately equal to 1 tonne (metric ton)]

6.14 (a)

Year	1760	1800	1840	1880	1920	1960	2000
actual concentration (ppm)	275	280	285	290	300	315	365
actual increase	–	5	10	15	25	45	85
percentage increase	–	1.8	3.6	5.4	9.0	16.4	30.9

(b) increasing atmospheric carbon dioxide coincides with steep increase in carbon emission from fossil fuel burning, but note earlier date given on graph for fossil fuel burning;

(c) carbon dioxide is a greenhouse gas, so contributes to greenhouse effect;

(d) sulphur dioxide; oxides of nitrogen;

6.15 (a)

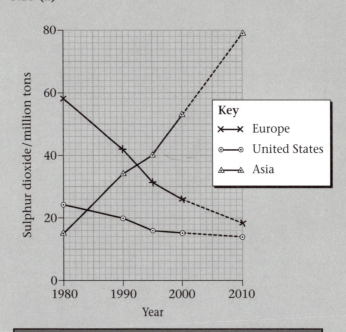

Examiner's comments

In plotting your graph, remember to label your axes, tags (on the axes) should go outside, the plotted points should be joined with straight lines and give a key.

(b) *Europe* – refer to legislation, e.g. has taken steps to reduce sulphur emission in line with treaty on acid rain (from 1979);
Asia – increase in population; increase in industrialisation, not involved in treaties / legislation / agreements to reduce emission;

6.16 organise as a flow chart, but essentials are:

$SO_2 \rightarrow$ oxidation $\rightarrow SO_4^{2-}$ ⟶ washed out as rain (H_2SO_4)
⟶ dry deposition as SO_2

$NO_x \rightarrow$ oxidation $\rightarrow NO_3^-$ ⟶ washed out as rain (HNO_3)
⟶ dry deposition as NO_x

6.17 carbon dioxide; oxides of nitrogen (NO_x);

6.18

Organisms		How affected by acid rain
aquatic organisms	phytoplankton	loss – so less material in suspension, clearer water in acidified lakes;
	fish	loss – less Ca^{2+}, leads to mobilisation of Al^{3+}, excess mucus on gills, difficulty in gas exchange, trout larvae may be trapped in eggs; etc.;
	other species	e.g. moss *Sphagnum* – vigorous growth, blankets other species;
terrestrial organisms	trees – conifers	low pH leads to mobilisation of Al ions, damages roots; less uptake of nutrient; waxy cuticle on leaves breaks down; loss of nutrient, trees weaker hence disease etc.; affects microorganisms in soil, slower decomposition, hence less nutrient;
	trees – deciduous	similar to conifers but effects usually less severe;
humans (health)		low pH releases ions of heavy metals in soil, or copper and lead from water pipes, may contaminate drinking water;

6.19 ~30 °C lower; you can work out for yourself how the lower temperature would affect your life. Imagine that instead of hot summer temperatures of ~30 °C, your 'hot' days were ~0 °C. Then list which crops might grow in these temperatures, or how you could make them grow. Compare present day tropics, or daytime temperatures in the Sahara Desert, with Arctic conditions to help you visualise the difference.

6.20 (gases of atmosphere) – equivalent to glass of the glasshouse; (UV radiation 'filtered out') absorbed by reactions between oxygen and ozone; (heat energy 'retained') by warm layer of the troposphere; ('greenhouse gases' accumulate) in the troposphere – absorb more heat, less reflected back to space;

6.21

Name of greenhouse gas	Origin or source
carbon dioxide	*respiration; combustion of fossil fuels; combustion of other organic materials (e.g. forest burning);*
chlorofluorocarbons (CFCs);	aerosols; refrigerators;
methane (CH_4);	guts of livestock (such as ruminants); anaerobic bacteria; rotting vegetation;
nitrous (and other) oxides (NO_x)	*combustion of fossil fuels; rapid decomposition (e.g. of humus);*
ozone	*internal combustion engines;*

order of effectiveness = carbon dioxide (× 1), methane (× 11), nitrous oxide (× 270), CFCs (× 3400+);

6.22 sharp rise in global temperatures coincides with increases in greenhouse gases (carbon dioxide, CFCs, methane, nitrogen oxides) – but this cannot *prove* cause and effect, i.e. that global warming is anthropogenic rather than a climatic fluctuation, similar to ones that have occurred before;

6.23 some examples – higher temperatures, higher yields of crops, different crops in (say) temperate regions, also affects weeds and pests; changed distribution of flora and fauna in natural ecosystems; altered weather patterns, affects rainfall, winds, etc.; warmer temperatures, less fuel used for heating; raising of sea levels (due to melting of ice caps and glaciers) – hence loss of coastal land and change of ecosystems, loss of agricultural land; etc.;

Examiner's comments

It is difficult to be precise as changes could be complex, diverse and widespread. Climatic modelling takes into account many variables, but predictions do not give all the answers. Make sure your answers are reasoned using biological principles you know about, rather than emotional and based on inadequate knowledge. Not all changes are bad!

6.24 reduce burning of fossil fuels; more efficient energy use; conservation of energy (insulation etc.); use of renewable energy sources; recycling / re-use of materials; look at use of energy for transport and travel; reforestation; legislation – encouragement and enforcement of agreements (for industrial and domestic fuel use); etc.;

6.25 *Greenhouse effect* – trapping of heat in lower atmosphere due to accumulation of certain 'greenhouse gases' (such as carbon dioxide, methane etc.) in upper atmosphere (see textbook for full explanation);

Global warming – rise in temperature on Earth, attributed mainly to enhanced greenhouse effect, often linked with anthropogenic activities;

6.26 (a) increases; e.g. anaerobic respiration (microorganisms) / breathing tubes to surface / haemoglobin as respiratory pigment etc.; high temperatures; high BOD (from many / population explosion of organisms demanding oxygen); low rate of photosynthesis (which would replenish oxygen); little movement or stagnation of the water; mechanical agitation of water, e.g. paddles, pumping over falls etc.; fast-moving mountain stream over rocky bed / eq.;

(b) increases; carbon dioxide in water forms carbonic acid – a weak acid, so high carbon dioxide means low pH; daily fluctuation related to light intensity and temperature changes;

during daylight, carbon dioxide used in photosynthesis; so pH becomes higher as day progresses; pH lower in darkness as carbon dioxide increases; higher rate photosynthesis with higher temperatures (during day); different seasons – higher rate photosynthesis with longer daylength (summer) and higher temperatures, so higher pH in summer and lower pH in winter;

(c) light intensity decreases with depth in water, soil / sediment;
organic particles / debris, microbes / algae / etc.;
reduce rate of photosynthesis;

6.27 *Oligotrophic* – low nutrient content;
Eutrophic – rich in nutrients;

Phytoplankton – free floating microscopic plants in water (mainly unicellular algae, including diatoms);
Zooplankton – microscopic animal life in water (mainly small crustaceans and fish larvae);

6.28 run-off of fertiliser applied to agricultural land; discharge of sewage (e.g. could be from boats on the water) / slurry from farm animals; discharge of detergents into the water (mainly phosphates);

6.29 (a) high nutrient encourages rapid growth and reproduction of algae and blue-green bacteria;

(b) rapid growth of population suddenly stops when certain nutrients become limiting; leads to massive death of the algae (rise to surface as a scum); bacterial decay of dead algae makes heavy demand on oxygen supply; water becomes

depleted of oxygen / anaerobic;

(c) reduces light penetration; reduces rate of photosynthesis for other aquatic plants in the water; leads to further depletion of oxygen (because not produced by photosynthesis);

6.30 sudden death of masses of fish, usually due to oxygen depletion; likely at night because of nocturnal respiration by algae (algal bloom); can occur at other times if heavy demand on oxygen linked with decay of dead algae; fish kills also due to toxins / thermal pollution / eq.;

6.31 toxins from certain algae; growth of bacteria such as *Clostridium botulinum* which flourish in anaerobic conditions (and produce toxins); scum of algae (living and dead) which clog gills of fish;

(b) algae – 'recovery';
bacteria – at point of sewage discharge;
Cladophora – 'active decomposition';
protozoa / protoctista – 'degradation' + 'active decomposition';
sewage fungi – 'degradation' + 'active decomposition';

(c) *Asellus* – peak in 'recovery', starts in 'active decomposition';
Chironomus – peaks (earlier) in 'recovery', starts (earlier) in 'active decomposition';
clean water fauna – peak in 'clean water', start at beginning of 'recovery';
Tubificidae – peak in (early) 'decomposition', start at point of sewage discharge;

6.32 (a)

Zones downstream (above and below discharge)	Oxygen	BOD	Dissolved solids Suspended solids	Ammonia	Nitrates	Phosphates
clean water	high	low	low low	low	low	low
degradation	falls	rapid rise	rapid rise rapid rise	rapid rise to high	drops	rises
active decomposition	falling	falls	falling falls rapidly	peaks then falls	rises	falls
recovery	rises	falls	falling to low at low level	falling	falling	falling
clean water	high	falls	low low	low	low	low

Practice questions

1 (a) (i) nitric acid; nitrous acid; carbonic acid; sulphurous acid; **[2]**

(ii) burning / eq. of fossil fuels or named example; production / eq. carbon dioxide / sulphur dioxide; nitrogen oxides from car exhausts / eq.; gas(es) combines with water to produce (dilute) acid; **[3]**

(b) (i) as concentration increases, root length decreases / eq.; growth in 1×10^{-3} molar causes an increase; a suitable comment about an anomalous result; **[2]**

(ii) both decrease; no shoot growth above 8×10^{-3} molar but roots do / eq.; shoots more sensitive / eq. / converse for roots; all concentrations decrease shoot growth; **[3]**

(iii) grow quickly / rapid germination; easy to grow / large numbers can be grown / easy to measure; easy to control other variables; easy to treat / whole plant can be exposed / eq.; **[2]**

(Total 12 marks)

2 phosphates / eq.; bacteria / appropriate named; biochemical / biological oxygen demand; increases; (sun)light;

(Total 5 marks)

Answers to assessment questions

Unit 1 Molecules and cells

1

Property / Role	Starch	Glycogen	Cellulose
constituent monomers	α-glucose	*α-glucose;*	*β-glucose;*
type of bonds present	*1,4 glycosidic for amylose;* *1,4 with 1,6 for amylopectin;*	*1,4 glycosidic with some 1,6 (more 1,6 than amylopectin);*	1,4 glycosidic
nature of polysaccharide chain	*amylose unbranched;* *amylopectin branched;*	very branched	*straight unbranched chains; hydrogen bonding between chains;*
role in living organisms	energy store in plants	*energy store in animals;*	*structural component of cell walls in plants;*

(Total 8 marks)

2 **(a)** active site; [1]

(b) substrate only fits into active site; substrate complementary to site; R groups interact / bond with substrate; [2]

(c) active site-directed inhibitor similar shape to substrate; binds to active site, preventing entry of substrate; non-active site-directed inhibitor binds elsewhere; changes shape of active site, preventing entry of substrate; rate of reaction decreases; [3]
(Total 6 marks)

3 cell wall; vacuole / tonoplast; chloroplasts / plastids; plasmodesmata; starch grains (any three of these in the first three spaces); prokaryotic / bacterial; tissue / mesophyll;
(Total 5 marks)

4 **(a)** movement of molecules across membranes against a concentration gradient / from a low concentration to a high concentration; requires metabolic energy / energy from ATP; reference to an example / Na / K pumps / eq.; [3]

(b) incorporation of enzymes in insoluble substances without losing catalytic activity; such as entrapment within polymer gels / adsorption on to charcoal / alginate beads / eq.; enzyme does not have to be separated from product / reference to ease of harvesting product; enzyme can be re-used / eq.; [3]

(Total 6 marks)

5 **(a) (i)** activity at 65 °C = 35 units, activity at 75 °C = 47 (units); percentage increase = (12 ÷ 35) × 100; 34.28 / 34.3 (%); [3]

(ii) increasing temperature increases kinetic energy / movement of molecules; increases chances of collision / formation of enzyme–substrate complexes; *therefore* activity increases; optimum at 75 °C; above optimum enzyme becoming denatured / eq.; so active site distorted / changed in shape / no longer complements substrate / eq.;

[4]

(iii) stable at high temperatures / less likely to be denatured; temperature control less important; can be used continually / ref. to high productivity / faster; reduced risk of contamination by other microorganisms;
[2]

(b) activity will decrease / eq.; heavy metal inhibitors / disrupt internal bonding / tertiary structure; [2]

(Total 11 marks)

6 The following points would gain credit:

any reference to DNA being a polynucleotide / eq.; nucleotides consist of deoxyribose, phosphate and a base; bases are adenine, guanine, cytosine and thymine; A and G are purines; C and T are pyrimidines; DNA is a double helix; phosphate–sugar linked to form strands / eq.; strands are antiparallel / run in opposite directions; A pairs with T and C pairs with G; hydrogen bonding between base pairs; replication semi-conservative explained; replication occurs during interphase; strands separate due to breaking of H bonds; new strand synthesised by DNA polymerase; complementary base pairing reference to give accurate copy; reference to unwinding / separation of the strands by helicase; 5' to 3' direction of synthesis by DNA polymerase; lagging / eq. strand sections joined by ligase;

(Total 10 marks)

Unit 2B Exchange, transport and reproduction

1 guard; potassium; more negative / lower; cell walls; open; **(Total 5 marks)**

2 **(a)** involvement of active transport; since all potassium ions are absorbed; against a concentration gradient; energy / ATP from respiration needed; **[3]**

(b) plant takes in water; reference to transpiration / water loss from the barley plants; **[2]**

(c) seedlings all kept at the same temperature; same light intensity; duration of the experiments the same; same humidity; plants all at same stage of growth / all same age / size / eq.; exclusion of light from roots to prevent algal growth; **[5]**

(d) root hair region / eq.; cortex; endodermis / pericycle; reference to the apoplast / symplast / vacuolar pathways as appropriate; **[3]**

(Total 13 marks)

3 **(a)** A = vascular bundle / vascular tissue / vein / xylem and phloem / eq.; B = epidermis / epidermal cell; **[2]**

(b) presence of hairs / eq. of epidermis / surface; reduce air movement / trap air / trap water vapour / humidity reference; thick cuticle; reduces evaporation / diffusion of water / water loss; leaf rolled / eq.; reduces external / exposed surface area / reduction in air movement / humidity reference; hinge cells; roll / eq.; leaf in dry conditions; stomata few / sunken in pits / on inner epidermis; reduced water loss through pores / eq.; **[4]**

(Total 6 marks)

4 **(a)** A = sino-atrial node / SA node; B = atrioventricular / AV node; **[2]**

(b) 0.16 – 0.04 = 0.12 seconds; **[1]**

(c) excitation spreads across atria so atria contract; excitation passes to the ventricles via the AV node; reference to non-conducting tissue between atria and ventricles; reference to time delay at the AV node / conduction to ventricles; excitation passes to the apex of ventricles; ventricles contract after atria / ventricular systole after atrial systole; **[4]**

(Total 7 marks)

5 **(a)** has an affinity for oxygen at high partial pressure / concentration / tension of oxygen / haemoglobin combines with oxygen; gives up oxygen (readily) at low partial pressure / low oxygen / high concentration of carbon dioxide; **[2]**

(b) (i) loading tension 10.8 kPa (allow from 10.6 to 10.8); unloading tension 3.5 kPa (allow from 3.4 to 3.6); answer 7.3 kPa; **[2]**

(ii) location – any body tissue / organ except lungs; reason – oxygen being used up / reference to respiration / respiring tissue; **[2]**

(c) curve shifts to the right; (increasing carbon dioxide makes) haemoglobin less efficient at taking up oxygen; haemoglobin has to be exposed to higher partial pressure / eq. oxygen in order to reach loading tension / loading tension higher; higher partial pressure / eq. of oxygen; so unloading tension higher / eq.; credit reference to some detail, e.g. hydrogen ion production inducing release of oxygen from haemoglobin; **[4]**

(d) has higher affinity for oxygen than mother's blood / haemoglobin; in order to be able to take up oxygen from mother's blood / haemoglobin via placenta; **[2]**

(e) in solution / dissolved in the plasma; as hydrogencarbonate / bicarbonate ions; associated with the haemoglobin / as carbaminohaemoglobin; **[3]**

(Total 15 marks)

6

Statement	First division of meiosis	Second division of meiosis
pairing of homologous chromosomes occurs	✔	✘
chromosomes consist of pairs of chromatids during prophase	✔	✔
chiasmata are formed	✔	✘
chromatids are separated	✘	✔
independent assortment of chromosomes occurs	✔	✘

(Total 5 marks)

7 insect-pollinated small quantities of pollen / more for wind-pollinated; pollen heavy / light for wind-pollinated; pollen rough / sticky / sculptured / smooth for wind-pollinated; scented flowers / no scent in wind-pollinated; nectar / no nectar in wind-pollinated; reference to the position of the stigma in insect-pollinated; position of the anthers in insect-pollinated; reference to the entry of insects and movement of flower parts / exposure of anthers; no petals in wind-pollinated; anthers outside the flower in wind-pollinated; loose attachment of anther to filament in wind-pollinated; feathery / large surface area of stigma in wind-pollinated; flowers above foliage in wind-pollinated; position of stigma in wind-pollinated;

(Total 10 marks)

Examiner's comments

Note the large number of marking points and that some of them are alternatives for wind or insect pollination, whereas others are stand-alone points. The best way to plan this answer would be to draw up a table of differences.

Unit 2H Exchange, transport and reproduction in humans

1 **(a)** increased ventilation due to less efficient gas exchange; reduced efficiency of gas exchange due to increased mucus; increased emphysema; formation of carboxyhaemoglobin; **[3]**

(b) nicotine can cause constriction of placental blood vessels; increased risk of spontaneous abortion; formation of carboxyhaemoglobin slows growth of fetus; lower birth weight; **[3]**

(Total 6 marks)

2 **(a) (i)** decreases; **[1]**

(ii) haemoglobin greater; so more oxgen carried; 12 % / 20 % more haemoglobin / some manipulation of data; 9 % / 14 % more oxygen carried / manipulation of data; **[3]**

(iii) greater lung capacity / increased cardiac output / stroke volume / more red blood corpuscles / increased diphosphoglycerate / hyperventilation / barrel-chested; **[1]**

(b) humidity; temperature; wind speed; solar radiation / ultraviolet light / eq.; **[3]**

(c) hyperventilation / breathlessness / eq.; increased cardiac output / eq.; pulmonary oedema / eq.; nausea / eq.; headaches / eq.; impaired mental activity / lethargy; **[3]**

(Total 11 marks)

Unit 3 Energy and the environment

1 **(a)** A = enamel ridge; B = dentine; C = cement; D = pulp cavity; **[4]**

(b) open root; allows continued growth of tooth as it is worn down; enamel ridge; provides hard grinding surface; large surface area; efficient grinding of plant material; **[4]**

(Total 8 marks)

2 **(a)**

$$\frac{1.8 \times 10^4 \times 100}{3 \times 10^6} \quad \left| \quad \frac{18{,}000 \times 100}{3{,}000{,}000} \quad \right| \quad \text{eq.; 0.6 \%;} \quad \textbf{[2]}$$

(b) light reflected from plants / from leaf / not absorbed by plant / leaf / eq.; energy used to evaporate (water) / heat plant / eq.; light transmitted / eq.; light of wrong wavelength; photosynthesis / biochemical processes inefficient; released by plant respiration; **[2]**

(c) 1800 kJ transferred to caterpillars, only 100 kJ to birds / calculation of loss / amount transferred (5.6 %); any reference to energy loss; reference to loss of energy / biomass as respiration / excretion / in movement / eq.; **[2]**

(Total 6 marks)

3 **(a)** penetrates the food source / anchors the organism on food; secretes digestive enzymes / eq. / soluble products of digestion; absorbed; **[3]**

(b) *Taenia* lives in another living organism / in gut of cow / eq.; food digested by host organism / eq.; so no need for digestive enzymes / soluble food absorbed directly / eq.; **[3]**

(c) in mutualism two organisms live in a partnership / named example; both contribute / both benefit / description; **[2]**

(Total 8 marks)

4 **(a)** known or stated volume of water sampled / description e.g. use a sweep net for a fixed distance; organisms sorted into trophic levels / consumer groups; organisms in each group weighed; several samples to obtain a mean; **[3]**

(b) (i) (391.2 ÷ 418) × 100;
= 93.6 %; **[2]**

(ii) respiration / loss as carbon dioxide; not all organisms eaten / death / losses to decomposers; not all that is eaten is digestible / egestion / faeces / excretion; **[2]**

(Total 7 marks)

5 **(a)** keep out oxygen / maintain anaerobic conditions / eq.; maintain constant temperature / eq.; prevent gas escaping; **[2]**

(b) (i) 0–10 days – bacteria produce carbon dioxide / organic acids; carbon dioxide / organic acids increase acidity / causes pH fall;

10–45 days – organic acids converted to methane; acids used up; causes pH increase; **[3]**

(ii) initial increase due to carbon dioxide; from bacteria / aerobic respiration; production falls due to less oxygen / anaerobic / pH fall / eq. / more organic acids; production rises due to methane production; suitable conditions for anaerobes / anaerobic respiration / different bacteria / eq.; as temperature rises / pH is low / changes; last few days – substrate depleted / no more organic acids, so production falls; **[3]**

(c) little or no sulphur in biogas / no sulphur dioxide produced (*accept converse if fossil fuels mentioned*); does not contribute to acid rain; renewable (biomass); uses waste products / eq. / readily available substrates; production cheaper / time quicker / less damaging; **[3]**

(Total 11 marks)

Index

Page numbers in italics indicate that information is contained in a question and its answer. You will need to look at both pages together.

For example the entry

radiation, heat transfer through 52(94)

indicates that you need to look at a question on page 52 and its answer on page 94. Note that in this instance the word radiation only appears on page 94, although information about radiation is on both pages.

abiotic factors 66(97)
absorption 3
acclimatisation
    altitude 56(96)
    climate 53(95)
acid rain
    gases which cause 77–8(101–2)
    sulphuric acid in 80–1(103)
acids
    amino acids see amino acids
    fatty acids 11
activation energy 10(88)
    enzyme action 8
active sites 9(87)
active transport 18(104)
    across cell membranes 13(88)
adaptations 31–2, 32–3(91–2)
    aquatic invertebrates 31, 32–3(92)
    humans 44
        of highlanders 56(96), 60(106)
        to climate 53(95)
    plants 32(91–2)
adipose tissue 53(94–5)
afforestation 74(100)
air pollution 73–4, 75
    acid rain 77–8(101-2)
        sulphuric acid in 80–1(103)
    greenhouse effect 78, 78(102)
algae 79(103)
alimentary canal 22(89)
    ruminants 63–4(97)
altitude see high altitude
amino acids 6(87)
    DNA specifies 3
    from enzyme catalysis 10(87)
    R groups 7
amylase, enzyme action of 10(87)
amylose see starch
anaphase 16(89)
animals
    as consumers 61, 61(97)
    invertebrates, aquatic 32–3(92)
        and oxygen concentration 31
    mammals
        ruminant guts 63–4(97)
        transport in 28–31
    polysaccharides in 5(86), 7(87)
    teeth and jaws 63(96), 81–2(106)
    see also organisms
antibodies 30(91)
antigens 30(91)
aquatic ecosystems
    energy flow in 67(98)
    oxygen in 31
        invertebrate adaptation 32–3(92)
    pollution in see water pollution
    pyramids of biomass
        collecting material 65, 67(98)
        trophic levels in 83(106)
        units used 64
aquatic organisms
    acid rain affects 78(102)
    invertebrates 31, 32–3(92)
arteries 30(90)
atmospheric pollution see air pollution
autotrophic nutrition 61, 63(96)

bacteria see Rhizobium
Benedict's reagent 4
biodiversity, loss of 74(100)
biogas 72(100), 83–4(106)
biological detergents 8
biological molecules 2–7
biomass 65

and energy crops 72(99)
    pyramids of see pyramids of biomass
biosphere 66(97)
Biuret reagent 4
blood 28–9
    capillaries 53(94–5)
    and dissociation curves 30(91), 42(105)
        in hypoxia 56(95)
    functions 29(90), 47
    in hypoxia 55(95), 56(95)
    tissue fluid 49(93)
blood cells 30–1(91)
    composition 28, 30(90)
    microscope work on 29, 48
blood vessels
    arteries and veins 30(90)
    capillaries 53(94–5)
body temperature 50–1(94)
    in cold conditions 56–7(96)
    skin regulates 53, 53(94)
Bohr effect 30(91)
bonds, protein structure and 7
breathing 20–1, 22(89)
    humans 45, 46–7(93)
breathing rates, measuring 46
buffer solutions 9
buildings, climate and 52(94)

calcium alginate 9
capillaries 53(94–5)
carbohydrates 2
    digestion 22(89)
    structure 11
carbon cycle 68, 68(98), 69(98)
carbon dioxide
    in aquatic habitats 79(102)
    atmospheric 76–7(101)
        in deforestation 74(100)
        as greenhouse gas 78(101)
    in blood 42(105)
        erythrocyte take-up 30(91)
    in hypoxia 55(95)
cardiac output in hypoxia 55(95)
carnivores 63(96)
catalysts, enzymes as 8
cells 10–13
    blood see blood cells
    cell cycle 14–15, 15–17(88-9)
    palisade cells 14, 18(104)
    stem cells 30–1(91)
    surface membranes 11, 12–13(88)
cellulose 17(104)
    digestion by ruminants 63–4(97)
    as polysaccharide 5(86), 7(87)
CFCs (chlorofluorocarbons) 78(102)
chemical reactions see condensation reactions;
    hydrolysis reactions; rate of reaction
chlorofluorocarbons (CFCs) 78(102)
chlorophyll 65
chloroplasts 13(88)
chromosomes 14
circulatory systems 28–9, 29–31(90–1)
    heart see heart
    in low oxygen 54–6(95)
    materials transported by 44, 47
climates
    cold see cold climates
    hot climates 52–3(94)
    see also extremes of temperature
clothing
    for high altitude 54(95)
    for hot dry climates 52(94)
    for hot humid climates 52(94)
coal 71(99)
cold climates
    body temperature in 56–7(96)
    cold stress in 52(94)
    at high altitude 54(95)
    visitors and natives in 53(95)
    see also hot climates
cold injury 52(94)
cold stress 52(94)
collecting material 65
    aquatic habitats 67(98), 83(106)
colorimeters 4
colour standards 4
communities 67(98)
condensation reactions 3–4
    of amino acids 6(87)

enzymes in 10(88)
    of monosaccharides 5(86)
conduction 51(94)
consumers 61, 66(97)
convection 51(94)
covalent bonds 7
cytokinesis 14

decomposers 67(98)
deforestation 73, 74, 74(100)
denaturation 7, 9(87), 10(88)
deoxyribonucleic acid see DNA
desertification 73, 74, 75(100–1)
detergents, biological 8
diet 21, 22–3(89–90)
    see also digestion; nutrition
differential white cell count 29, 48
diffusion
    and exchange surfaces 20, 44
    facilitated 13(88)
diffusion gradients 20, 44
digestion 3, 21, 22–3(89–90)
    ruminant 63–4(97)
    see also diet; nutrition
dipeptides 6(87)
diploid chromosomes 14
disaccharides 3, 5(86)
    see also lactose; maltose; sucrose
dissociation curves 30(91), 42(105)
    in hypoxia 56(95)
DNA (deoxyribonucleic acid)
    condensation reactions form 4
    in organelles 13(88)
    structure 6–7(87)
        bases in 3
DNA polymerase 10(88)
DNA replication 7(87), 16–17(89), 19(104)
double circulatory system 30(90)
dry climates 52(94)

ecology, human 50–7, 50–3(94–5)
    extremes of temperature 50–3(93–4)
    high altitude 53, 54–6(95)
ecosystems 64–5, 66–7(97–8)
    aquatic see aquatic ecosystems
electrocardiograms 49(93)
electrovalent bonds (ionic bonds) 7
energy
    in chemical reactions 10(88)
    in ecosystems 64–7, 66–7(98)
        aquatic 67(98)
        food chains and webs 61, 82(106)
energy crops 72(99)
energy resources 70–1, 71–3(99–100)
    fossil fuels see fossil fuels
enhanced greenhouse effect 78
environment
    adaptations to 31–3
        humans 44, 53(95), 56(96), 60(106)
    exchanges with 20–3
        humans 44–7
    human influences on 73–81
enzymes 7–10, 9–10(87–8)
    immobilised see immobilised enzymes
    inhibitors 10(87), 18(104)
    and temperature 18–19(104)
epidermis, stomatal counts on 26
epithelium 46(93)
erosion
    and deforestation 74(100)
    and desertification 75(100)
erythrocytes (red blood cells) 30(90)
    carbon dioxide take-up 30(91)
eutrophication 79(103)
evaporation 52(94)
exchange surfaces 20, 44
exchanges with the environment 20–3, 22–3(89–90)
    humans 44–6, 46–7(93)
excretory products 45, 47
exercise, investigating 48
extremes of temperature 50, 50–3(94–5)
    cold conditions 56–7(96)
eyepiece graticules 12

facilitated diffusion 13(88)
fatty acids 11
feeding see nutrition
fertilisers, water quality and 75, 79(103)

fetal haemoglobin 30, 42(105)
fish kill 79(103)
flowering plants
    grasses 34, 43(105)
    reproduction 34, 35, 37(92)
        meiosis 38(93)
        pollination see pollination
    stomata 22(89), 39(105)
    transport 24–6, 26(90), 27–8(91)
        mineral uptake 25, 39–40(105)
    xeromorphic features 31
follicle stimulating hormone 38(92)
food chains 64, 65, 67(98)
    energy in 82(106)
food modification, pectinases in 8
food webs 65, 67(98)
fossil fuels 71(99)
    and acid rain 77(101)
    and the carbon cycle 69(98)
    carbon emissions from 76(101)
frostbite 52(94)
fructose
    and enzymes 10(87)
    high-fructose corn syrup 9
    as reducing sugar 4
fruit juice, increasing yield of 9
fuels 70–1, 71–3(99–100)
    fossil see fossil fuels
fungi (Rhizopus) 63(97), 83(106)

galactose, enzymes and 10(87)
gamete formation 34, 35
    in humans 36(92), 58
gas exchange 20–1, 22(89)
    in humans 45, 46–7(93)
        in hypoxia 55(95)
        and smoking in pregnancy 47(93),
            60(106)
gasohol 72(99), 73(100)
genes 3
genetics
    gamete fusion in 35
    mitosis in 14
germination, microscopic examination of 36
global warming 78(101–2)
glucose
    and enzymes 10(87)
    as monosaccharide 4(86)
    as reducing sugar 4
glucose isomerase 9
glycogen 6(86), 17(104)
    as polysaccharide 5(86), 7(87)
glycosides 11
glycosidic bonds 5(86), 11
grasses 34, 43(105)
graticules, eyepiece 12
greenhouse effect 78, 78(101–2)
greenhouse gases 78(102)
    reducing emission of 78(102)
guts see alimentary canal

habitats 66(97)
    see also ecosystems
haemoglobin
    dissociation curves 30(91), 42(105)
        in hypoxia 56(95)
    in hypoxia 55(95), 56(95)
hair erector muscles 53(94–5)
hanging drop preparations 36
HCG see human chorionic gonadotrophin
heart 29(90)
    electrocardiograms 49(93)
    heart beat 41(105)
        in hypoxia 55(95)
    pacemakers 49(93)
heat collapse 52(94)
heat exhaustion 52(94)
heat stress 52(94)
heat stroke 52(94)
heat transfer mechanisms 51–2(94)
herbivores 63(96), 81–2(106)
    see also ruminants
hexoses 4(86)
    see also fructose; galactose; glucose
high altitude 53, 54–6(95–6), 60(106)
high-fructose corn syrup 9
holozoic nutrition 62, 63(96)
hormones
    reproductive 38(92)
        in menstrual cycle 39(93)
        in pregnancy 59–60(96)
    transport of 47

hot climates 52–3(94)
    see also cold climates
human chorionic gonadotrophin (HCG) 59–60(96)
human ecology 50–7, 50–3(94–5)
    extremes of temperature 50–3(93–4)
    high altitude 53, 54–6(95)
human reproduction 34–5, 57–60
    gamete formation 36(92)
    pregnancy 59–60(96)
        smoking in 47(93), 60(106)
    reproductive systems 37–8(92)
humans
    body temperature see body temperature
    and breathing 45, 46–7(93)
    and the carbon cycle 69(98)
    energy resources used by 70–3
    and the environment
        damage to 73–81
        exchanges with 44–7
    transport in 28–31, 44, 47–9
        see also blood; blood cells; heart
humid climates 52(94)
hydrogen bonds 7
hydrolysis reactions 3
    enzymes in 10(88)
hydrophilic regions 11
hydrophobic regions 11
hydrophytes 32(91)
hyperventilation 55(95)
hypothermia 52(94), 57(95)
hypoxia 54–6(95)

immobilised enzymes 10(88), 18(104)
    glucose isomerase 9
    lactase 8
inhibition, enzyme 10(87), 18(104)
insect-pollination 34
    flower features for 37(92), 43(105)
invertebrates, aquatic 32–3(92)
    and oxygen concentration 31
iodine, starch test for 4
ionic bonds (electrovalent bonds) 7
irrigation 75(100)

jaws 63(96)

lactase
    enzyme action 10(87), 10(88)
    immobilised form 8
lactose
    as disaccharide 5(86)
    enzymes act on 10(87), 10(88)
landfill gas 72(99–100)
leaves
    as organs 11
    palisade cells 14, 18(104)
    stomata see stomata
    tissues in 13(88)
    of xerophytes 40(105)
leucocytes (white blood cells)
    differential count 29, 48
    as phagocytes 30(90), 31(91)
    types 30(90)
light see sunlight
lipids 3
liver 11, 14
lungs 23(90)
    inflation of 22(89)
luteinising hormone 38(92)

maltase, enzyme action of 10(87)
maltose
    as disaccharide 5(86)
    enzymes act on 10(87)
mammals
    ruminant guts 63–4(97)
    teeth and jaws 63(96), 81–2(106)
    transport in 28–9, 29–31(90–1)
    see also humans
maximum rate of reaction 9(87), 10(87)
meiosis 34, 35
    in flowering plants 38(93)
    in humans 57–8
        in gamete formation 36(92)
    stages of 43(105)
membranes, cell 11, 12–13(88)
menstrual cycle 39(93)
metabolic waste 45, 47
metaphase 16(89)
methane, greenhouse effect and 78(102)
microorganisms
    in mutualistic nutrition 63(96)

in nitrogen cycle 68(98)
    and sewage discharge 80(103)
    see also Rhizobium
microscopes 12
    blood cells in 29, 48
    meiosis in 36
    mitosis in 15
    ovary sections in 59
    root tip squash in 15
    testis sections in 36, 59
milk production 58
mineral uptake 25, 39–40(105)
mitochondria 13(88)
mitosis 14, 16–17(89)
    microscopic examination of 15
    mitotic index 15
molecules, biological 2–7
monomers 3–4
    of carbohydrate 11
    in disaccharides 5(86)
    see also monosaccharides
mononucleotides
    condensation reactions 4
    in DNA and RNA 3
monosaccharides 4(86)
    see also fructose; galactose; glucose
mountain sickness 56(96), 60(106)
mutualistic nutrition 63(96), 83(106)

native residents
    adaptations 44
        to climate 53(95)
        to high altitude 56(96), 60(106)
    at high altitude 54(95)
natural gas 71(99)
NFFO (Non-Fossil Fuel Obligation) 72
nitrates 70(99)
    and sewage discharge 80(103)
nitrogen, oxides of, as greenhouse gas 78(102)
nitrogen cycle 68, 69–70(99)
    microorganisms in 68(98)
    Rhizobium in 63(96)
Non-Fossil Fuel Obligation (NFFO) 72
non-reducing sugars, test for 4
nucleic acids 3
    see also DNA; RNA
numbers, pyramids of 66(97)
nutrients
    in deforestation 74(100)
    movement in plants 25
    recycling see recycling nutrients
    and water pollution 79–80(103)
nutrition modes 62–4, 62–4(96–7)
    see also diet; digestion

oestrogen 38(92)
oil 71(99)
oil immersion lenses 29, 48
optimum pH, enzyme 9(87), 10(87)
organelles 12–13(88)
organisms
    acid rain affects 78(102)
    investigating oxygen uptake 21
    and sewage discharge 80(103)
    see also animals; plants
organs 11
ovary, microscope sections 59
oxygen
    in aquatic habitats 79(102)
        concentrations 31, 32
        and invertebrates 32–3(92)
        and sewage discharge 80(103)
    atmospheric, in deforestation 74(100)
    at high altitude 60(106)
        hypoxia 54–6(95)
    investigating uptake 21
oxygen dissociation curves 30(91), 42(105)
    in hypoxia 56(95)
oxytocin
    in milk production 58
    secretion and effects 38(92)
ozone 78(102)

pacemakers 49(93)
palisade cells 14, 18(104)
parasites 63(96)
    Taenia 63(97), 83(106)
partial pressure, dissociation curves and 42(105)
pectin, enzymes act on 10(87), 10(88)
pectinase
    enzyme action of 10(87), 10(88)
    in food production 8, 9

peptides, from enzyme catalysis *10(87)*
pH, enzymes and 7, 9, *9(87)*, *10(87)*
phagocytes *30(91)*, *31(91)*
phloem 25, *26(90)*
phosphates
    in phospholipids 11
    and sewage discharge *80(103)*
phospholipids *5(86)*
    in membranes 11
    in organelles *13(88)*
photosynthesis 61, 64, *66(97)*
    and energy flow *67(98)*
plants
    adaptations *32(91–2)*
    grasses 34, *43(105)*
    pollination 34
        flower features *37(92)*, *43(105)*
    polysaccharides in *5(86)*, *7(87)*
    reproduction 34, 35, *37(92)*
        meiosis *38(93)*
        pollination *see* pollination
    stomata *22(89)*, *39(105)*
    transport 24–6, *26(90)*, *27–8(91)*
        mineral uptake 25, *39–40(105)*
    xerophytes 31, *32(91)*, *40(105)*
pollen grains, microscopic examination of 36
pollination 34
    flower features *37(92)*, *43(105)*
pollution 73–4, 75, *76–81(100–3)*
polymers 3–4
    see also polypeptides; polysaccharides
polynucleotides 4
polypeptides 3
    synthesis *6(87)*
polysaccharides *5(86)*
    condensation reactions 3
    in living organisms *7(87)*
    monomers in *17(104)*
    see also cellulose; glycogen; starch
potometers 26
pregnancy
    HCG in *59–60(96)*
    smoking in *47(93)*, *60(106)*
pressure, partial, and dissociation curves *42(105)*
prickly heat *52(94)*
primary consumers *66(98)*
producers 61, *67(98)*
    in food webs *66(98)*
progesterone *38(92)*, *60(96)*
prolactin *38(92)*, 58
prophase *16(89)*
protease
    in detergents 8
    enzyme action *10(87)*
proteins 3
    as enzymes *see* enzymes
    enzymes act on *10(87)*
    and phospholipid membranes 11
    test for 4
pulse rate, investigating 48
pyramids of biomass
    collecting material 65, *67(98)*
    trophic levels in *83(106)*
    units used 64
pyramids of numbers *66(97)*

R groups 7
    functions *6(87)*
radiation, heat transfer through *52(93)*
rate of reaction 9
    maximum *9(87)*, *10(87)*
rate of respiration 21
receptors *53(94)*
recycling nutrients 67, *68–70(98-9)*
    carbon cycle 68, *68(98)*, *69(98)*
    nitrogen *see* nitrogen cycle
    water cycle *68(98)*

red blood cells (erythrocytes) *30(90)*
    carbon dioxide take-up *30(91)*
reducing sugars, test for 4
reduction division 34, 35, 58
reforestation *74(100)*
renewable resources *71–3(99-100)*
replication, DNA *7(87)*, *16–17(89)*, *19(104)*
reproduction 34–9, *36–9(92–3)*
    human *see* human reproduction
    plants 34, 35, *37(92)*
        meiosis *38(93)*
        pollination *see* pollination
reproductive systems 20
    human *37–8(92)*
residual groups *see* R groups
respirometers 21
    using *22(89)*
Rhizobium (bacterium) *63(97)*
Rhizopus (fungus) *63(97)*, *83(106)*
ribonucleic acid *see* RNA
rivers, deforestation and *74(100)*
RNA (ribonucleic acid) 3, 4
    in organelles *13(88)*
root tip squash, microscopic examination 15
roots *26(90)*
ruminants, guts *63–4(96–7)*

salinisation *75(100)*
saprobiontic nutrition *63(96)*
secondary consumers *66(98)*
sewage discharge 75, *80(103)*, *81(103)*
sexual reproduction 34–9, *36–9(92–3)*
    human *see* human reproduction
    plants 34, 35, *37(92)*
        meiosis *38(93)*
        pollination *see* pollination
    reproductive systems 20
        human *37–8(92)*
size, organism, and transport 24
skin 53, *53(94-5)*
smoking, pregnancy and *47(93)*, *60(106)*
soil, deforestation and *74(100)*
solar radiation
    at high altitude *54(95)*
    see also sunlight
spermiogenesis *36(92)*
spirometers 46, *46(93)*
squash preparations 15, 36
starch *17(104)*
    enzymes act on *10(87)*
    as polysaccharide *5(86)*, *7(87)*
    test for 4
stem cells *30–1(91)*
stomata
    counting 26
    opening of *22(89)*, *39(105)*
    in transpiration *27–8(91)*
structure, protein 7
substrates, enzyme *10(87-8)*
sucrase, enzyme action of *10(87)*
sucrose
    as disaccharide *5(86)*
    enzymes act on *10(87)*
    as non-reducing sugar 4
sugars
    tests for 4
    see also disaccharides; monosaccharides
sulphur dioxide, acid rain and *77(101)*
sulphuric acid, acid rain and *80–1(103)*
sunburn *52(94)*
sunlight
    in ecosystems 65
        aquatic *67(98)*
    and water pollution *79(102)*
    see also solar radiation
surface area/volume ratio 20, 24, *26(90)*
    humans 44

surfactants *23(90)*
sweat glands *53(94-5)*

Taenia (tapeworms) *63(97)*, *83(106)*
teeth *63(96)*, *81–2(106)*
telophase *16(89)*
temperature
    body temperature *50–1(94)*
        in cold conditions *55–6(96)*
        skin regulates 53, *53(94-5)*
    and enzymes 7, 9
        commercial preparation *18–19(104)*
    extremes of 50–3, *50–3(94-5)*
        cold conditions *56–7(96)*
testes sections 36, 59
tests, food 4
thermogenesis *53(95)*
thermoregulation *53(95)*
tissue fluid *49(93)*
tissues 11
transpiration *26(90)*, *27–8(91)*
    factors affecting *32(92)*
    investigating with potometers 26
transport 20, *39–42(105)*
    across cell membranes *13(88)*
    in humans 28–31, 44, 47–9
        see also blood; blood cells; heart
    in mammals 28–31
    in plants 24–6, *26(90)*, *27–8(91)*
        mineral uptake 25, *39–40(105)*
trench foot *52(94)*
triglycerides *5(86)*
trophic levels 65
    energy loss at *66(98)*
turnover numbers *9(87)*

urine in hypoxia *56(95)*
UV radiation, greenhouse effect and *78(102)*

vascular tissue *26(91)*
vasoconstriction *53(95)*
vasodilation *53(95)*
veins *30(90)*
ventilation *22(89)*
    and smoking in pregnancy *47(93)*
    see also breathing
villi *23(90)*
visitors 44
    climate acclimatisation *53(95)*
    mountain sickness *56(96)*
vital capacity, measuring 46

water 2
    and exchange surfaces 20, 44
    plants
        transpiration *see* transpiration
        uptake and transport 24
        xerophyte adaptations 31, *32(91)*, *40(105)*
water cycle *68(98)*
water pollution 73–4
    by fertilisers 75, *79(102)*
    by sewage discharge 75, *80(103)*, *81(103)*
white blood cells *see* leucocytes
wind-pollination 34
    flower features for *37(92)*, *43(105)*

xerophytes 31, *32(91)*, *40(105)*
xylem 24, *26(90)*

zygotes 35